Results and Problems in Cell Differentiation

Series Editor:
W. Hennig

39

Springer

Berlin
Heidelberg
New York
Barcelona
Hong Kong
London
Milan
Paris
Tokyo

Christine F. Hohmann (Ed.)

Cortical Development

From Specification to Differentiation

With 28 Figures, 5 in Color

Springer

Dr. Christine Hohmann
Department of Biology
Morgan State University
Cold Spring Lane and Hillen Road
Baltimore, MD 21251
USA

ISSN 0080-1844
ISBN 3-540-43436-4 Springer-Verlag Berlin Heidelberg New York

Library of Congress Cataloging-in-Publication Data
Cortical development: from specification to differentiation / Christine F. Hohmann (Ed.).
p. cm. – (Results and problems in cell differentiation; 39)
Includes bibliographical references and index.
ISBN 3540434364 (alk. paper)
1. Neocortex – Growth. 2. Developmental neurophysiology. I. Hohmann, Christine F., 1955–
II. Series.
QH607 .R4 vol. 39
[QP383.12]
571.8′35 s – dc21
[573.8′619]

Springer-Verlag Berlin Heidelberg New York
a member of BertelsmannSpringer Science+Business Media GmbH
http://www.springer.de

Cover design: Meta Design, Berlin
Typesetting: SNP Best-set Typesetter Ltd., Hong Kong
SPIN 10780806 39/3130 – 5 4 3 2 1 0 – Printed on acid-free paper

Preface

The cerebral neocortex, unique to mammals, is regarded as the prerequisite for higher cognitive function and is the structure most closely associated with the idea of the "mind". Expansion of mental capacity between mammals is most typically associated with an evolutionary increase in neocortical volume that culminates in the intricately folded configuration of sulci and gyri so characteristic of the primate cerebral cortex. Yet, the basic unit structure and fundamental connectivity of cortex appears to have been preserved from the smooth cortex of the mouse or rat to the highly convoluted cortical mantle of the human that, if stretched out as a sheet, would be large enough to wrap the entire human brain multiple times. The basic similarity in structure and function has made it possible to conduct studies in the relatively simple cortices of rat or mouse and have the results pertain to the understanding of the primate, including human, cortex.

The neocortex is an intriguing structure for the study of cell differentiation. Its dozens of neuronal cell types and small handful of different glial types have their origin in a pseudostratified germinal epithelium lining the ventricular surface of the forebrain. In its mature form, neocortex is a six-layered structure; five of its layers contain multiple different but characteristic neuronal types with the sixth occupied by neuronal processes. Various glial cells are dispersed throughout all six layers. Its precise stratification is essential for the proper function of neocortex. Postmitotic, undifferentiated cells must migrate to their proper position within the developing cortex. Once arrived in their correct cortical stratum, neurons must elaborate their differentiated characteristics while simultaneously establishing appropriate synaptic connections with their axonal targets and the afferent axons of other neurons of cortical as well as diencephalic and brainstem origin. Moreover, specific morphological and neurochemical relationships between neurons and glia must evolve during the course of morphogenesis. Mishaps during any of these steps can result in cascades of developmental alterations, leading to functional changes, as has been learned from the study of mouse mutations and the pathologies of human developmental disorders.

While the first glimpse into the complexities of neocortical development was allowed us by Ramon y Cajal, a century ago, the study of cortical morphogenesis came into its own in the early 1960s. Studies by Angevine, Racik, Sidman, Bayer and colleagues described the time line of "birth" among cortical cells, their basic modes of migration along the scaffolding of radial glia cells and elucidated the fundamental "inside-out" pattern of cortical development whereby the earliest wave of neurons forms the deepest cortical layer and subsequent waves of neurons will migrate past earlier ones to become situated in more superficial positions. Mutant mice with various abnormalities in cortical formation helped to reveal a plethora of cell–cell and cell–matrix interactions instrumental in the proper assembly of the neocortex. Many of the cell surface and diffusible signals involved in altered migration and subsequent abnormal differentiation in these mutants have since been identified, while others are still under study.

Along with the description of these basic rules of neocortical ontogeny, research began to reveal the fundamental plasticity of the developing neocortex. Like no other brain structure, cortical morphogenesis has shown itself amenable to molding by sensory inputs from the periphery as well as motor feedback. Such plasticity was demonstrated early on in the kitten visual system by experiments that would eventually help D. Hubel and T. Wiesel earn the Nobel Prize for Medicine. Many studies since have shown that use-dependent plasticity is a universal feature of neocortex. The elaboration of cortical neuronal fine structure (dendrites, their spines, synapses), and the resulting perceptual and behavioral outcomes, is shaped by activity patterns in afferent axons onto cortical neurons. These activity patterns, in turn, are representative of patterns of peripheral sensory inputs and/or motor feedback.

Thus, understanding the ontogeny of cerebral cortex requires us to understand not just the step-by-step "readout" of a developmental program, but also the manner in which this program is designed to interact with environmental influences. While this might appear to some a daunting task, perhaps to be abandoned for simpler systems, the reward of understanding cortical differentiation on a functional level will be to understand how a thinking and learning, cognitive brain is assembled and continues to evolve. In other words, understanding the cortex is to understand how we understand! On a biomedical level, such knowledge eventually will enable us to prevent and/or remedy numerous disorders of the brain and "mind".

Contemporary research into cortical differentiation has at its disposal the new tools of molecular biology that, combined with a systems approach to the study of the nervous system, will eventually reveal the full picture of the functional assembly of the cerebral cortex. Exciting new insights into cortical development are uncovered on both the molecular and systems level with increasing frequency and have recently been augmented by computational approaches as well. In this first volume of a two-volume issue on cortical development, we will concentrate on "early" events in ontogeny, namely, issues related to cortical cell generation/specification, migration, and early differentiation. Many of the questions before us in this arena are of a molecular nature. The second volume will concentrate, to a larger degree, on issues involved in cortical plasticity and associated function and, thus, will feature models using a systems approach.

The chapters assembled here have been selected to show a cross section of what I personally regard as the most pertinent and exciting areas in current research on early cortical ontogeny. I have made an attempt to generate a diversity of voices, although, for reasons of practical limitations, not every area of research could be sampled and not every opinion represented. I will leave it to readers of this volume to "dig deeper" into areas of interest and uncover the debates and controversies that may exist, as they do in any field of scientific inquiry.

Chronologically, the earliest questions we encounter in cortical morphogenesis are: How does the germinal epithelium of the ventricular lining control the generation of the many different kinds of neurons and glia emerging during the days or weeks of cortical cell generation? Are selected populations of precursor cells generated first, which later will give rise to limited populations of differentiated cells or is the germinal epithelium composed of pluripotent stem cells responding to timed signals with differentiated responses? In any case, what are the molecular signals that interact with such stem cell or precursor cell populations to specify their future differentiation?

The chapters by Richard Nowakowski and colleagues and Mark Mehler address these questions from very different angles. R. Nowakowski et al. look at patterns of postmitotic cell emergence and the changes in cell cycle regulation that accompany their generation as a source for controlling differences in cell populations. Mehler reviews and investigates a number of different growth factors and like molecules in the specification of postmitotic neuronal and glial lineages. As different as these approaches seem, they are intimately connected as cell cycle activity is clearly subject to regulation by many of the growth factors at issue. Bernd Sutor introduces us to another perspective on neuronal genesis, the role of Gap junctions which exist between clusters of cells of the germinal epithelium and allow for communications between these cells concerning internal (e.g., cell cycle) as well as external (e.g., receptor-mediated) signals. Studies discussed in this chapter furthermore address the early stages of cortical neuronal differentiation and the role of Gap junctions as well as the emergence of synaptic communication in these differentiation steps.

There has been considerable speculation for some time that differentiating cortical neurons and glia depend on each other, as well as on certain afferent axons, to generate their appropriate differentiated morphologies and circuits. The chapter by Juliano and Gierdalski revisits neuronal-glia interactions during the migratory phase that sets up cortical lamination. Her work suggests that focal disruptions of neuronal genesis will alter molecular signaling between the remaining glial and neuronal populations in ways that suggest a profoundly complex information exchange between the maturing cellular elements of the emerging neocortex. Juliano et al.'s data moreover indicate that normal afferent thalamocortical ingrowth can proceed despite profoundly altered cortical morphogenesis.

Bernd Sutor proceeds to gives us some insights into how direct neuronal communication, via Gap junctions and/or excitatory and inhibitory chemical synapses, may shape cortical differentiation. Kimberly McAllister, on the other hand, illuminates the role of diffusible neurotrophins in the differentiation of neuronal processes and elaboration of synaptic contacts. Her chapter allows us a glimpse of the intricate relationship between these neurotrophic factors and synaptic activity in shaping cortical ontogeny as well as use-dependent plasticity.

Katrin Andreasson and Walter Kaufmann take us inside the differentiating cortical neuron by exploring the intracellular signaling pathways and "immediate, early" gene activation that occur in response to both synaptic and neurotrophic signaling in the developing cortex. Here, we can begin to see how synaptic activity mediated and neurotrophin mediated intracellular responses might converge on signaling chains leading to gene activation. Once more, as in the chapter by McAllister, parallels between the regulation of morphogenesis and plasticity become evident.

While much of the current focus in cortical development is directed towards molecular and cellular studies, many questions continue to require a look at the "larger picture". We have known for centuries that the neocortex is subdivided into clearly delineated functional areas. Research during the last few decades has detailed the synaptic circuitry of several such functional areas, in particular, the different sensory areas, and has shown that substantial differences exist, for instance, in how visual and auditory synaptic processing occurs. There has been a long-standing debate as to whether such structural and functional differences are the result of intrinsic cortical signals or the consequence of information conveyed to cortex by afferents of different sensory modalities. Several laboratories have recently identified genes that are expressed at appropriate developmental times, marking specific cortical fields. The implication has been that the products of these genes guide afferent target finding and the establishment of functional cortical areas. However, studies presented here by Alvin Lyckman and Mriganka Sur present a powerful argument for the important role of sensory afferents in shaping the functional innervation of neocortex; simultaneously, their data concede that some secondary specifications of cortical circuitry are apparently intrinsically programmed. The independence of thalamocortical afferents from intrinsic cortical differentiation signals does echo the findings communicated earlier in this volume by Juliano and Gierdalski.

I am concluding this volume with a second chapter by Mark Mehler. I share with him a profound interest in the clinical relevance of current research into cortical morphogenesis and plasticity. This final chapter reviews recent data concerning the origin and morphogenetic regulation of cortical cells from the ventral telencephalic eminence, an area viewed in the past as giving rise to deeper forebrain structures. Most importantly, however, this chapter makes the provocative claim that disruptions of early events in cortical cell generation, migration, and differentiation may be at the root of pathogenesis in neurodegenerative diseases by creating latent vulnerabilities. Degenerative disorders then arise in later life in response to otherwise nonlethal stressors of cell physiology that expose the underlying weakness.

In conclusion, I hope that this volume will update the initiated reader in regards to the current "hot spots" regarding studies of early cortical development while simultaneously raising interest among developmental biologists not currently familiar with this fascinating area of research. Increasingly, successful inquiry in the neurosciences depends on a strongly interdisciplinary approach. I believe this volume shows that, as we diversify in our research questions, we simultaneously, collectively, begin to approach the answers that will reveal the entire complicated mosaic of events.

Christine F. Hohmann
Department of Biology, Morgan State University,
Baltimore, Maryland, USA

Contents

Influence of Radial Glia and Cajal-Retzius Cells in Neuronal Migration
M. Gierdalski and S.L. Juliano

Neurotrophins and Cortical Development
A.K. McAllister

Population Dynamics During Cell Proliferation and Neuronogenesis in the Developing Murine Neocortex

RICHARD S. NOWAKOWSKI[1], VERNE S. CAVINESS JR[2], TAKAO TAKAHASHI[3], and NANCY L. HAYES[1]

Summary. During the development of the neocortex, cell proliferation occurs in two specialized zones adjacent to the lateral ventricle. One of these zones, the ventricular zone, produces most of the neurons of the neocortex. The proliferating population that resides in the ventricular zone is a pseudostratified ventricular epithelium (PVE) that looks uniform in routine histological preparations, but is, in fact, an active and dynamically changing population. In the mouse, over the course of a 6-day period, the PVE produces approximately 95% of the neurons of the adult neocortex. During this time, the cell cycle of the PVE population lengthens from about 8 h to over 18 h and the progenitor population passes through a total of 11 cell cycles. This 6-day, 11-cell cycle period comprises the "neuronogenetic interval" (NI). At each passage through the cell cycle, the proportion of daughter cells that exit the cell cycle (Q cells) increases from 0 at the onset of the NI to 1 at the end of the NI. The proportion of daughter cells that re-enter the cell cycle (P cells) changes in a complementary fashion from 1 at the onset of the NI to 0 at the end of the NI. This set of systematic changes in the cell cycle and the output from the proliferative population of the PVE allows a quantitative and mathematical treatment of the expansion of the PVE and the growth of the cortical plate that nicely accounts for the observed expansion and growth of the developing neocortex. In addition, we show that the cells produced during a 2-h window of development during specific cell cycles reside in a specific set of laminae in the adult cortex, but that the distributions of the output from consecutive cell cycles overlap. These dynamic events occur in all areas of the PVE underlying the neocortex, but there is a gradient of maturation that begins in the rostrolateral neocortex near the striatotelencephalic junction and which spreads across the surface of the neocortex over a period of 24–36 h. The presence of the gradient across the hemisphere is a possible source of positional information that could be exploited during development to establish the areal borders that characterize the adult neocortex.

The neocortex of mouse, and indeed of all mammals, has a six-layered organization. Moreover, there are strikingly similar numbers of neurons per unit area in different species (Rockel et al. 1974, 1980) albeit with some variation in different cytoarchitectonic areas (Beaulieu and Colonnier 1989; Beaulieu 1993). In contrast, however, the surface area of the neocortex is remarkably variable and accounts for most of the differences in neuron number in different species; for example, the mouse neocortex has a surface area that is about 1/400 of the surface area of the monkey neocortex (Haug 1987). These basic

[1] Department of Neuroscience and Cell Biology, UMDNJ-Robert Wood Johnson Medical School, 675 Hoes Lane, Piscataway, New Jersey 08854, USA
[2] Pediatric Neurology Service, Massachusetts General Hospital, 32 Fruit St, V-Burnham 901, Boston, Massachusetts 02114, USA
[3] Department of Pediatrics, Keio University School of Medicine, 35 Shinanomachi, Shinjuku-ku, Tokyo 160, Japan

Results and Problems in Cell Differentiation, Vol. 39
C.F. Hohmann (Ed.): Cortical Development

facts illustrate that both neuron number and the surface area of the neocortex are critical determinants of the adult structure; it follows that the regulation of neuron number and surface area must be an important feature of development. The regulation of neuron number per se must be under the control of only two developmental processes: i.e., cell proliferation and neuron death. In this chapter we will describe the dynamic changes that occur in the proliferating, migrating, and differentiating populations during neocortical development and illustrate how both the neuron number in the developing neocortex, as well as its surface area, are regulated by basic cell biological processes.

During development, most neurons destined for the neocortex arise from a pseudostratified epithelium that lines the lateral ventricles (i.e., a pseudostratified ventricular epithelium, PVE, see Table 1). For the most part, the cells of the PVE are located in the ventricular zone (VZ; Boulder Committee 1970). During late development a secondary proliferative population (SPP) appears in the location of the subventricular zone (SVZ). The PVE and SPP overlap somewhat at the border between the VZ and the SVZ. The PVE itself is relatively uniform in appearance both across the surface of the ventricles and as development proceeds; it approximately doubles in width early in the neuronogenetic interval but increases dramatically in surface area. Despite its uniform appearance, however, the behavior of this apparently homogeneous population of proliferating cells is not only greatly complex but changes systematically as development proceeds. At various stages during development, the output from the PVE comprises the cells of the secondary proliferative population and neurons and glia of the neocortex, and, thus, the proliferating cells of the PVE meet the definitions of both stem and progenitor cells (see Table 1 for definitions). It is not clear, however, whether the PVE cells are all stem cells, all progenitor cells, or some mixture of cells with multiple potentials. For this

Table 1. Nomenclature used

Stem cells	Cells that can produce neurons, glia, progenitor cells and also more stem cells.
Progenitor cells	Cells that can produce one lineage (e.g., neurons or glia) and more progenitor cells. In some systems the distinction between progenitor and stem cells may be a matter of degree of "stemness" (Blau et al. 2001).
Neurogenesis	The production of the cells of the nervous system (both neurons and glia).
Neuronogenesis	The production of neurons.
Neuronogenetic interval (NI)	The period of time during which neurons arising in the PVE become permanently postproliferative.
Pseudostratified ventricular epithelium (PVE)	Population of proliferating cells that lines the ventricles of the brain; the PVE is, for the most part, co-extensive with the VZ.
Ventricular zone (VZ)	Cytoarchitectonically defined layer that is adjacent to the ventricles of the brain.

reason, we shall refer to them as stem/progenitor cells, as PVE cells or as proliferating cells, whichever is appropriate for the context. However, as we shall see, the major output of the neocortical PVE is the neurons of the neocortex; thus, it seems likely that the majority of the PVE cells are neuronal progenitor cells and that their major function is neuronogenesis.

The principal behavior of a proliferating population is cell division. This involves passage through the cell cycle beginning at the end of one mitotic division and concluding with another mitotic cell division. With respect to the proliferative (as opposed to the histogenetic) fate of the daughter cells, there are a priori only three types of cell divisions (Fig. 1): symmetric, non-terminal cell division; symmetric, terminal; and asymmetric. Between each mitosis, the cells of the proliferating population pass through four distinct phases: G1, S, G2, and M. The S-phase of the cell cycle is the DNA synthetic phase during which the DNA content of the nucleus is replicated. This allows the markers of DNA synthesis to be used to define the S-phase, to study proliferating cells and to determine the kinetics of the movement of the proliferating cells through the cell cycle. The most widely used S-phase markers are thymidine analogs, in particular tritiated thymidine (^{3}H-TdR) and halogenated uridine, usually bromodeoxyuridine (BUdR), although iododeoxyuridine (IUdR) and chlorodeoxyuridine (CUdR) have also been used. ^{3}H-TdR has been available since 1959 (for a review, see Sidman 1970); generally, the tritium is in the methyl group so that the demethylated form (i.e., uridine) does not lead to incorporation into RNA. The incorporation of ^{3}H-TdR into cells is detected autoradiographically. The halogenated uridines are detected by immunohistochemistry using monoclonal antibodies to single-stranded DNA (Gratzner

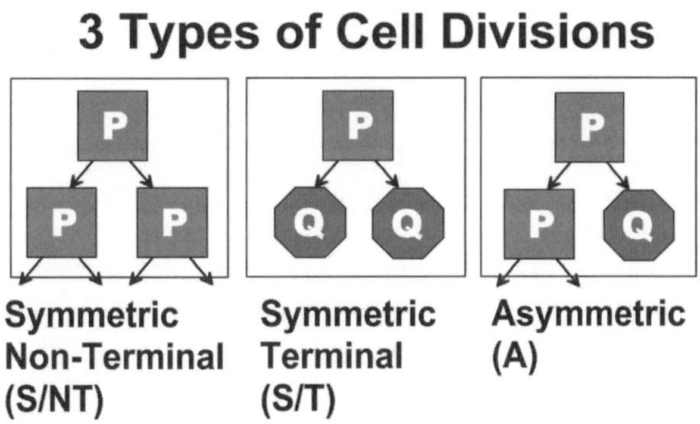

Fig. 1. With respect to the decision to re-enter the cell cycle, there are only three types of cell divisions. (1) A proliferating cell can divide to make two daughter cells, both of which re-enter the cell cycle and continue to proliferate (P cells); this is a symmetrical, non-terminal cell division. (2) A proliferating cell can divide to make two daughter cells, neither of which re-enters the cell cycle (Q cells); this is a symmetrical, terminal cell division. (3) A proliferating cell can divide to make two daughter cells of which only one re-enters the cell cycle (P cell) and the other does not (Q cell); this is an asymmetrical cell division

1982). In the developing nervous system, the incorporation of either ^{3}H-TdR or BUdR does not seem to affect adversely cell proliferation, migration or differentiation of neurons (Miller and Nowakowski 1988; Nowakowski et al. 1989), and, when detection methods that have similar sensitivity are used, the effective loading and clearance times are similar (Hayes and Nowakowski 2000). The loading time, i.e., the time needed to reach detectability, for both ^{3}H-TdR and BUdR is short, about 10 min; thus, only a short survival (e.g., 0.5 h) is needed to label S-phase cells (Takahashi et al. 1992; Hayes and Nowakowski 2000). Both markers remain at a sufficiently high concentration that rapid labeling of DNA continues for about 30–40 min (Hayes and Nowakowski 2000), which is consistent with the measured declines in the plasma levels (Nowakowski and Rakic 1974), but continued incorporation at low levels persists for 4–8 h after a single injection of either marker (Hayes and Nowakowski 2000). After this time, the levels of the markers apparently fall to a point where even continuous exposure of the proliferating cells during the entire S-phase is insufficient to produce enough incorporation to reach detection (Hayes and Nowakowski 2000).

The availability of S-phase markers provides a convenient point-of-entry for the study of the dynamics of cell proliferation. As described above, a single injection of an S-phase marker will label all the cells in the S-phase; the proliferating cells in the other phases of the cell cycle remain unlabeled (Fig. 2). Thus, if the tissue is collected after a short period of time (e.g., 0.5 h), a snapshot view of the location and number of cells in the S-phase is obtained. Assuming that the cells are randomly distributed in the cell cycle, the proportion of the proliferating population that will be detected is proportional to the length of the S-phase (Ts) divided by the length of the complete cell cycle (Tc) or Ts/Tc. However, if sufficient time is allowed to elapse, some of the labeled cells will exit S and enter G2, and other cells will enter S from G1 (for details, see Hayes and Nowakowski 2000). In the developing neocortex, a 2-h period approximates the length of the G2 + M phases (Takahashi et al. 1995; Hayes and Nowakowski 2000), so, for approximately 2.5 h after exposure to the label, all of the labeled nuclei will be distributed in S, G2 or M. If at this time (i.e., 2 h after injection of the first tracer) a second S-phase marker is given, a fraction of the proliferating cells proportional to the elapsed time (i.e., 2/Tc) will exit the S-phase and enter G2 (Fig. 2). At the same time, some cells will enter S from G1; these cells will not be labeled by the first marker but will be labeled by the second. The cells that remained in S during the 2-h interinjection interval will become double-labeled with the second marker (C and D in diagram on the right). This paradigm will effectively divide the proliferating population into four different types of cells, based on labeling characteristics. (1) The cells that exited S during the interinjection interval will be labeled only with the first S-phase marker. (2) The cells that entered S during the time between the injections and which did not get labeled by the first marker will be labeled only with the second S-phase marker. (3) Cells that were in S at the time of both injections will be labeled by *both* markers. (4) Proliferating cells that were

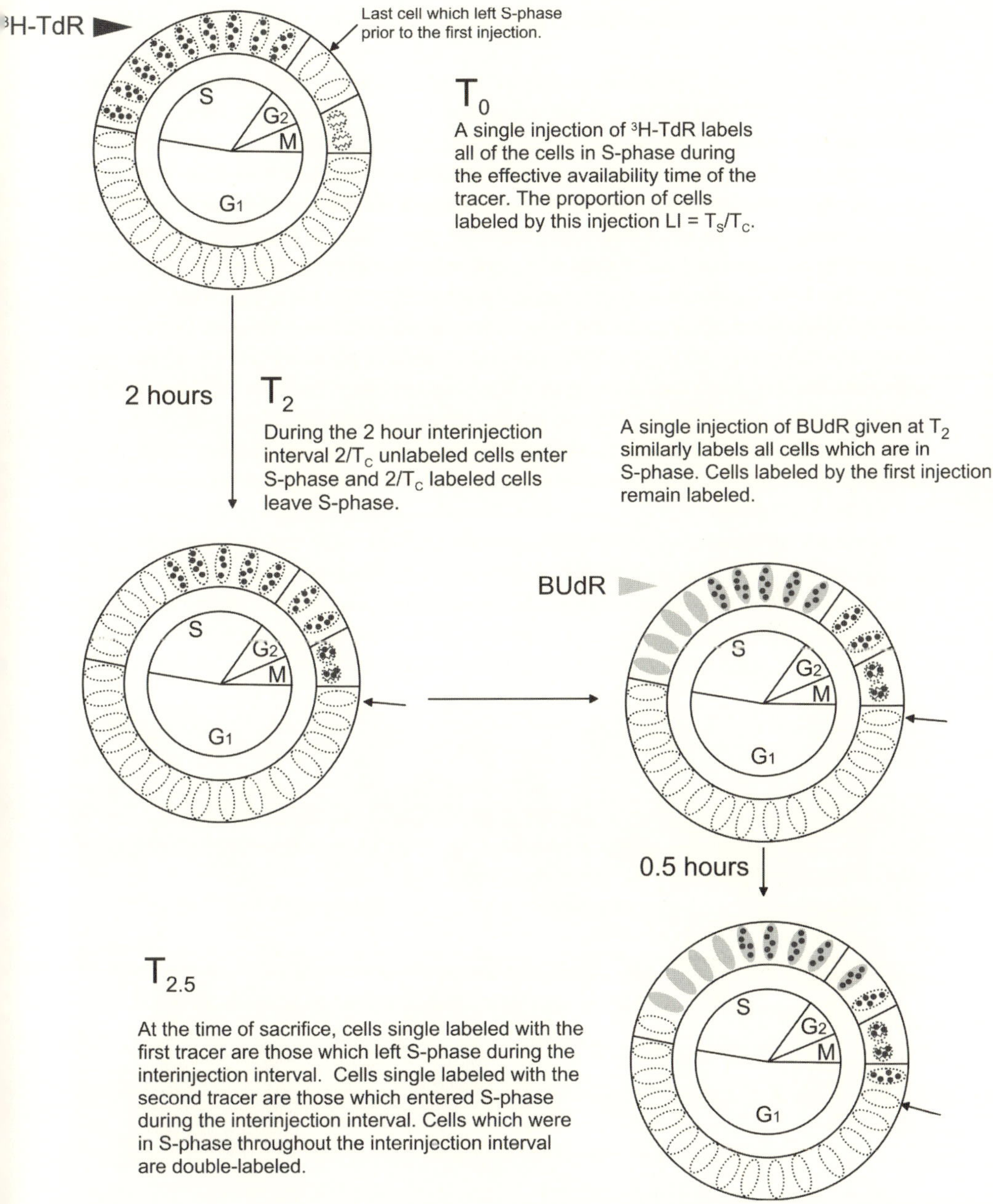

Fig. 2. Labeling paradigm used to identify cohorts of cells in specific phases of the cell cycle or that are born during a narrow window of time. The paradigm depends on the specificity of the S-phase labels, tritiated thymidine (^{3}H-TdR) or bromodeoxyuridine (BUdR). Either S-phase label is incorporated into cells synthesizing DNA at the time of the exposure, but cells in other phases of the cell cycle are unlabeled. By injecting the two tracers at different times, in this case with a separation of 2 h, cells that are labeled with only one tracer, with both tracers, or with neither tracer are produced

not at S at any time during the experiment will not be labeled with either S-phase marker.

An illustration of the labeling properties of the two commonly used S-phase markers, ^{3}H-TdR and BUdR, is shown in the figure on the right. When the two markers are given simultaneously, all of the labeled cells are marked by both markers. In addition, virtually all of the labeled cells are found in the outer part of the proliferative zone in what is called the S-phase zone (Takahashi et al. 1993). With an interinjection interval of 0.5 h, a few single-labeled cells of each type can be found (Fig. 3A, arrow and crossed arrow). In contrast, when the two markers are given 2 h apart (Fig. 3B), then the four types of cells as described above can be found in significant proportions (Hayes and Nowakowski 2000). What is additionally and strikingly evident in the two photographs in Fig. 3 is that, during the 2 h between the injection of the ^{3}H-TdR and the injection of the BudR, the nuclei of the cells labeled only with the ^{3}H-TdR have moved out of the S-phase zone towards the ventricular surface. By varying the time between the two injections, the nuclear movements can be monitored in detail (Hayes and Nowakowski 2000). Importantly, neither the movements of the nuclei nor the progression of the cells through the cell cycle as assessed by the changes in the proportions of various types of labeled cells are related to the sequence of injection of the two S-phase markers, i.e., indis-

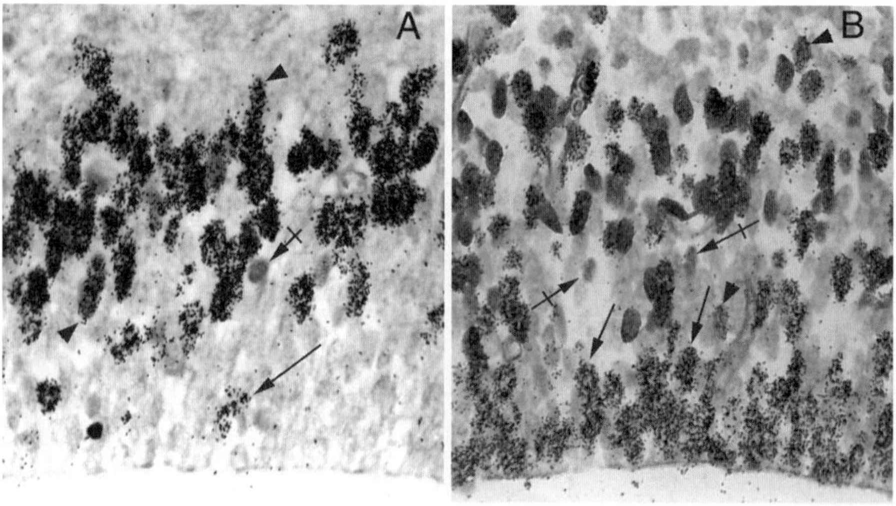

Fig. 3. Labeling pattern in the ventricular zone (VZ) of an E14 mouse that received ^{3}H-TdR followed either **A** 0.5 h or **B** 2.0 h later by BUdR and killed 0.5 h after the BUdR injection. Some cells labeled only with ^{3}H-TdR are indicated by an *arrow*; cells labeled only with BUdR are indicated by a *crossed arrow*. The cells labeled by BUdR define the S-phase, and they are located in the S-phase zone in the outer half of the VZ. The ^{3}H-TdR-only labeled cells have left the S-phase during the period between the two injections. The 2-h period (**B**) is long enough for many of the nuclei of the ^{3}H-TdR-only labeled cells to move towards the ventricular surface where they will divide

tinguishable results are obtained regardless of which marker is given first (Hayes and Nowakowski 2000).

The movement of the nuclei of the proliferating cells of the PVE during the cell cycle is called interkinetic nuclear migration. This phenomenon was first observed by Sauer (1936) and later confirmed by ^{3}H-TdR autoradiography (Sidman et al. 1959). More recently, we have shown that both the direction and speed of the nuclear movements are well correlated with the transition from one phase of the cell cycle to another (Takahashi et al. 1993; Hayes and Nowakowski 2000). Thus, as summarized in Fig. 4, as the cell cycle begins at the start of G1, nuclei are located at the ventricular surface. During G1, the nuclei move abventricularly towards the S-phase zone. The nuclei of cells entering S, in S and just leaving S are all distributed throughout the S-phase zone (Hayes and Nowakowski 2000) indicating that, during the S-phase, the nuclei have no net movement. However, upon exit from S, the nuclei move rapidly towards the ventricular surface. This adventricular movement is completed in less than 1 h and is, thus, 4–8× more rapid than the abventricular nuclear movement that occurs during G1. Mitosis occurs at the ventricular

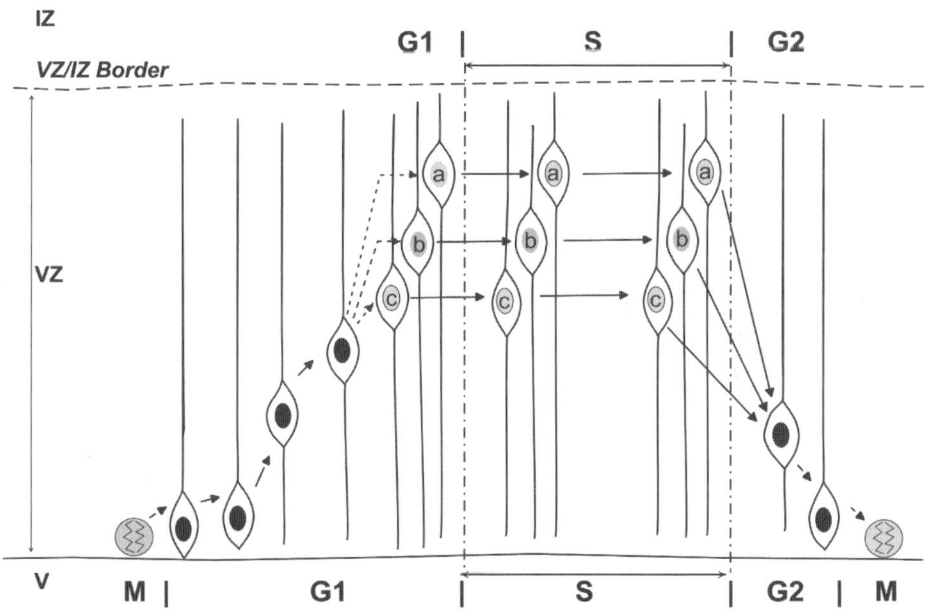

Fig. 4. Schematic diagram of the interkinetic movement of the nuclei of the cells of the PVE. During the cell cycle, both the direction and rate of movement of the nuclei are correlated with the phase of the cell cycle. Following mitosis, the nucleus of the PVE cells moves away from the ventricular surface. In the outer half of the VZ it enters the S-phase zone, where the nuclei labeled *a*, *b* and *c* represent nuclei distributed throughout the thickness of the S-phase zone in the outer half of the VZ. During S, the nuclei do not seem to move, but as they finish S and enter G2, they move rapidly back to the ventricular surface

surface. Interestingly, the orientation of the cleavage varies (Chenn and McConnell 1995; Chenn et al. 1998); it has been suggested that this variation is related to the decision to leave the cell cycle and to become postmitotic.

The amount of time required for a single cell to pass through one cell cycle, i.e., from the beginning of one G1 to the beginning of the next G1, varies systematically during the development of the neocortex (Fig. 5). At the time of the production of the first neurons in the mouse, neocortex measurements of the cell cycle using cumulative labeling with BUdR show that the total length of the cell cycle (Tc) is about 8 h, with an S-phase of about 3 h, G2 + M of about 2 h, and G1 of about 3 h (Takahashi et al. 1995). As development proceeds, the cell cycle lengthens until, at the end of the period of neuron production, Tc reaches ~18 h; S and G2 + M remain approximately constant at 3–4 and 2 h, respectively, and, thus, virtually all of the change in Tc is due to a increase in the length of G1 (T_{G1}; Takahashi et al. 1995). The increase in T_{G1} is dramatic, from 3 h early in development to almost 12 h late in development. As a result of the lengthening of the cell cycle, there are different numbers of cell cycles on each of the 6 days of production of neurons for the neocortex. We have defined the neuronogenetic interval (NI) as the period of time during which neurons arising in the PVE become permanently postproliferative. On the first day of the NI (E11), the cell cycle starts at 8 h and lengthens to over 10 h; thus, there is time for approximately 2.5 cell cycles. On the sixth and last day of the

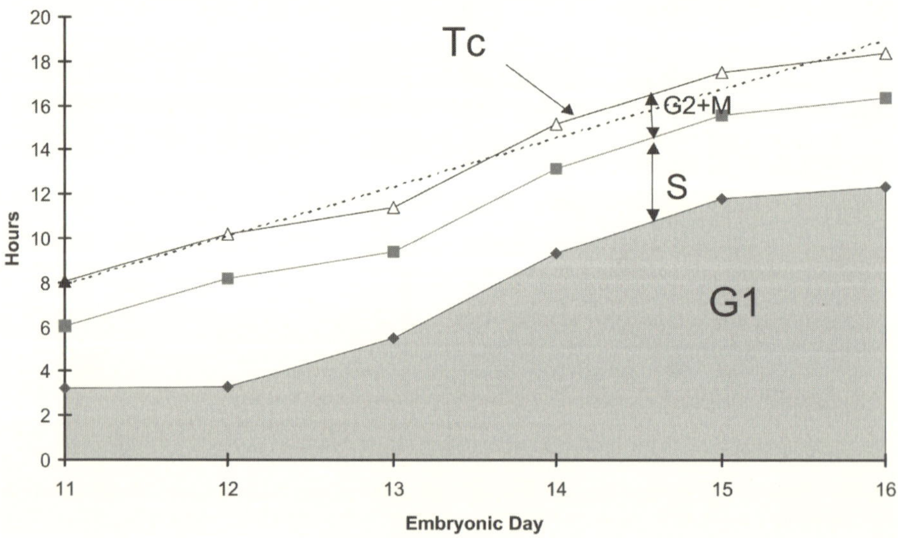

Fig. 5. The cell cycle lengthens systematically over the course of the 6-day period during which neurons are produced. At the onset of E11, the cell cycle is ~8 h, and by the end of the E16 it is over 18 h. During this time the length of G2 + M and S do not change systematically, and, hence, most of the lengthening is within G1

NI (E16), the cell cycle is over 18h, and, thus, there is sufficient time to complete only just over 1 cell cycle. By integration under the linear fit to the Tc (Fig. 5), we have determined that there is sufficient time for 11 cell cycles during the entire 6-day period (Takahashi et al. 1995). These determinations of the cell cycle lengths in the developing neocortex (Takahashi et al. 1995) were made using the cumulative labeling method (Nowakowski et al. 1989). With cumulative labeling methods the percentage of labeled cells rises linearly if the population is proliferatively homogenous, i.e., the length of the cell cycle and the S-phase are similar for all cells. Thus, from these measurements, we have estimated that the PVE of the mouse is 80–90% homogeneous (Takahashi et al. 1995) with cell cycle parameters within 10% of the mean (Nowakowski et al. 1989). This means, of course, that as many as 10–20% of the cells might have cell cycle parameters outside of this range. The cumulative labeling method gives an estimate for a maximum value of Tc because it is derived from the detection of an inflection point in the slope of the rising LI. This inflection point corresponds to the time required to label the entire proliferative population and occurs when the last (or slowest cycling) proliferating cell that was not labeled by the first injection enters the S-phase and becomes labeled (see Nowakowski et al. 1989). In contrast, an alternative method for measuring cell cycle lengths, the percent labeled mitosis method, gives an estimate of the minimum value of Tc because it detects the time required for the first (or fastest cycling) proliferating cell to transit the entire cell cycle and enter the M-phase for a second time (Kauffman 1968; Hoshino et al. 1973; Hamilton and Dobbin 1983a,b). The difference between the maximum and minimum estimates of Tc is an estimate of the range in Tc for the slowest versus fastest cycling cells. We used both methods to identify the range of the cell cycle lengths for the neocortical PVE and found that approximately 99% of the cells have a cell cycle within 5–7% of the mean (Cai et al. 1997). This means that if a population of proliferating cells with either a longer or a shorter cell cycle is present in the PVE, it comprises only about 1% of the total.

By definition, the beginning of the NI coincides with the first cell cycle during which neurons are produced. Thus, for the first cell cycle of the NI and for each of the subsequent cell cycles, to a total of 11, some of the daughter cells of the proliferating population exit the cell cycle (Fig. 6). The daughter cells that exit from the cell cycle are called "Q" cells, signifying that they are proliferatively *q*uiescent (or that they *q*uit the cell cycle). The daughter cells that remain in the cell cycle are called "P" cells because they re-enter the S-phase and, hence, continue to *p*roliferate. We also use "P" and "Q" to refer to the fraction of daughter cells that make each of these two choices. If, for the moment, we ignore the possibility of cell death within the proliferative population (considered in more detail below), it is clear that *all* of the daughter cells must select either a P or a Q fate and, hence, the proportions P and Q must add up to 1 (or $P + Q = 1$). Examination of the "old" tritiated thymidine birthday literature (e.g., Angevine and Sidman 1961; Caviness and Sidman 1973; Rakic 1974; Nowakowski and Rakic 1975; Stanfield and Cowan 1979) shows that in

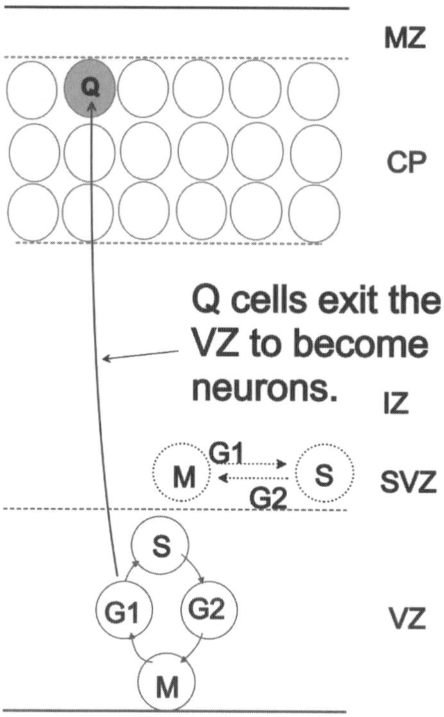

Fig. 6. Schematic diagram of the cell cycle in the VZ, where the PVE cells reside, and in the SVZ, where the SPP cells reside, showing that interkinetic nuclear movements occur for the PVE cells but not for the SPP cells. Also, during G1, some cells fail to pass through the restriction checkpoint and, instead of re-entering S, they migrate to the cortical plate as "Q" cells

various cortical structures different numbers of neurons are born on each of the various days of development. This means that P and Q must change dynamically as development proceeds. Thus, we decided that it was necessary to determine P and Q for each of the 6 days of development. From first principles, it seems clear that, prior to the onset of neuron production, P must be 1 and Q must be 0. Similarly, at the end of the NI, in order to account for the disappearance and involution of the PVE at the end of the NI, P must be decreased to 0 and Q must increase to 1. This means that during the NI P decreases from 1 to 0, and Q increases from 0 to 1.

In order to measure the P and Q proportions and determine the pathway that each takes from 1→0 or 0→1, respectively, we set up a double-labeling scheme to measure Q (Fig. 7) at the intervening stages of development. This double-labeling paradigm is a two-step process. In the first step, a 2-h cohort is set up by a procedure similar to that already described above (Fig. 2) using a single injection of ^{3}H-TdR followed 2 h later by an injection of BUdR. This leaves the cells that left the S-phase during the 2-h interinjection interval labeled *only* with ^{3}H-TdR. The goal is now to follow these ^{3}H-TdR-only labeled cells to determine what proportion of them exit the cell cycle as Q cells vs what proportion of them re-enter the cell cycle as P cells. To achieve this, animals

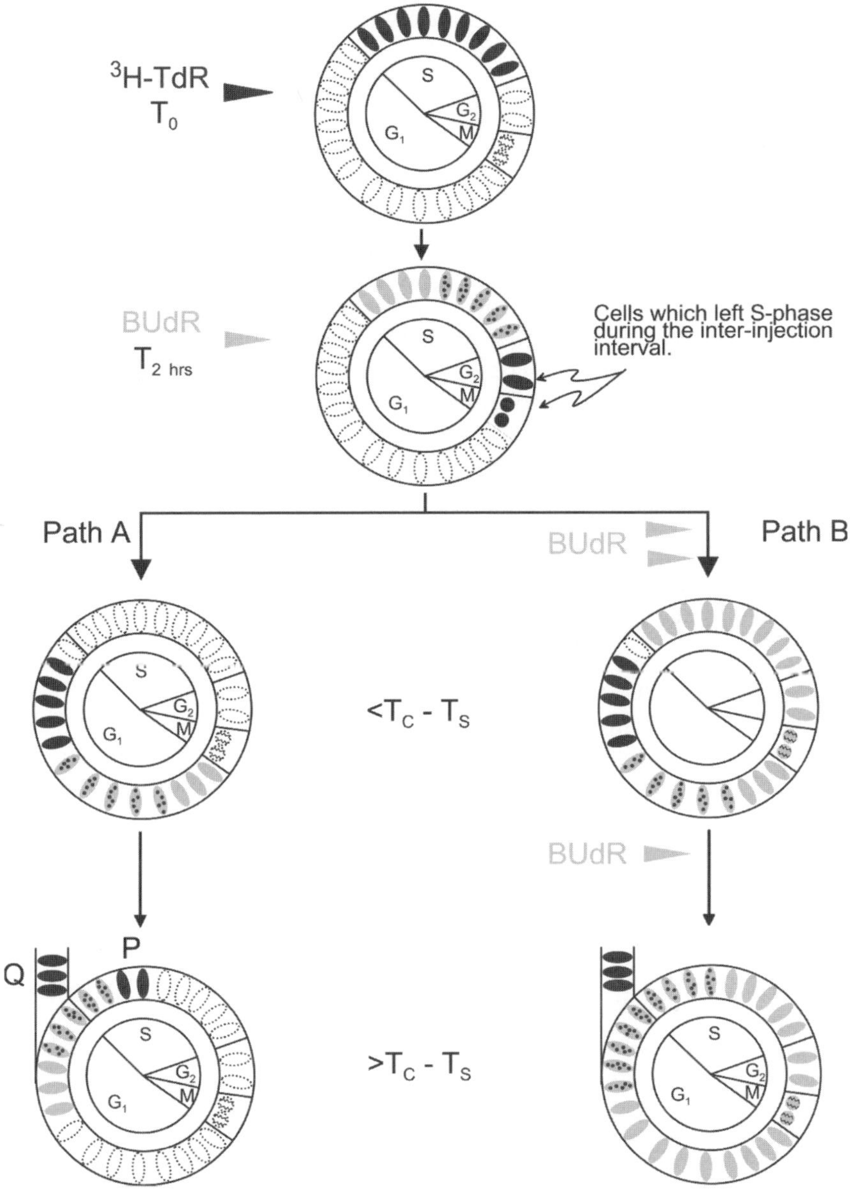

Fig. 7. Labeling paradigm used to measure Q and P in the developing neocortex. Initially, a 2-h cohort of ^{3}H-TdR-only labeled cells is established using the methods detailed in Fig. 2. Then, the animals are divided into groups. In *path A*, no additional label is given. At the end of a period that is greater than the length of the cell cycle (*Tc*) minus the length of the S-phase (*Ts*), the fetuses are harvested and processed for autoradiography and immunohistochemistry. The ^{3}H-TdR-only labeled cells are counted to get the total of Q + P cells in the 2-h cohort. In path B, BUdR injections are given repeatedly to label any cells that re-enter S. After a period of time > Tc–Ts, the fetuses are harvested and processed in this case to show the number of Q cells (i.e., those that did not get relabeled by re-entry into S)

with labeled 2-h cohorts were treated in two ways. One group (path A in Fig. 7) was allowed to survive for a period of time that is greater than Tc–Ts for that age (as determined above). The total number of ^{3}H-TdR-only labeled cells is a measure of the sum of *both* P and Q populations (or P + Q). The second group (path B in Fig. 7) was also allowed to survive for a period of time that is greater than Tc–Ts for that age, *but*, during the survival time, additional injections of BUdR were given at x-h intervals so that the P cells of the ^{3}H-TdR-only labeled cohort that re-enter the S-phase become double-labeled and are no longer distinguishable as part of the cohort. This means that the cells that are single-labeled with ^{3}H-TdR must be members of the 2-h cohort that have exited the cell cycle during the experimental period Tc–Ts, i.e., they are the Q cells. Thus, the number of Q cells is determined by counting the single-labeled ^{3}H-TdR cells in this second group of animals. Since the number of single-labeled ^{3}H-TdR cells in the other group of animals is the total number of cells in the 2-h cohort, or Q + P, the ratio of the number of ^{3}H-TdR cells in group 2 to the number in group 1 (Q/Q + P) is the Q fraction, and the P fraction is obtained by simple subtraction, i.e. P = 1–Q.

Measurements of Q made on each of the days of the NI in both dorsomedial and rostrolateral cortex (Miyama et al. 1997) are shown in Fig. 8. In Fig. 8 the abscissa shows the 11 cell cycles of the NI, and the data for each of the embryonic days is plotted at its appropriate proportional position on this 11 cell cycle scale. In addition to the measurements (shown as diamonds or triangles), the least-squares non-linear fit to the data points for each cortical region is shown. It is striking that the data and the least-squares fits obtained for the two cortical areas are essentially identical; moreover, the curves obtained do not differ significantly if only the actual data points are used or if they are "forced" mathematically to pass through Q = 0 at CC = 0 and Q = 1 at

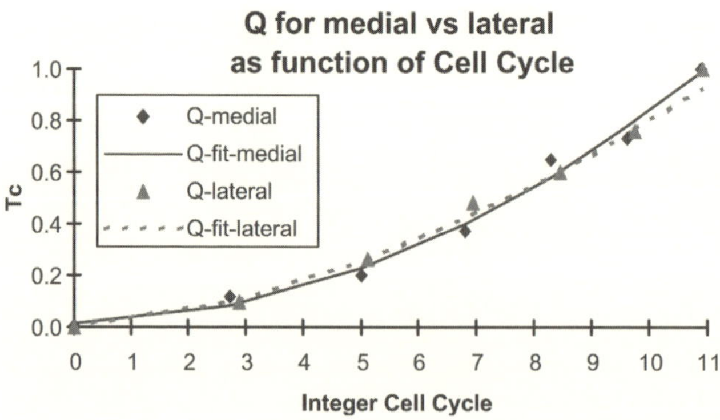

Fig. 8. During the neuronogenetic interval Q increases systematically and monotonically as a function of the number of elapsed cell cycles. The changes in both medial and lateral neocortex follow the same sequence when they are plotted by cell cycle number from the onset of neuron production despite the fact that the same events are occurring with a ~24-h separation

CC = 11. This indicates two things. First, it means that the progression of Q in two distant parts of the PVE that produce different cortical areas is indistinguishable. Second, it confirms the three starting assumptions, i.e., that the onset of the NI is E11 in the dorsomedial cortex and E10 in the rostrolateral cortex, that Q becomes greater than 0 at these times, and that Q reaches 1 at the end of the NI.

The path of Q = 0→Q = 1 increases monotonically; P is the complement of Q, and, thus, the path of P = 1→P = 0 decreases monotonically (Fig. 9). The P and Q curves intersect at P = Q = 0.5, which is between cell cycle 7 and 8. This divides the NI into two qualitatively different periods. During the first period, when Q < 0.5 and P > 0.5, the proliferative population expands. This is because, on average, more than 50% of the daughter cells produced re-enter the S-phase. For example, if Q = 0.3 and P = 0.7, then, for every 100 proliferating cells present at the beginning of the cell cycle, there will be 140 cells that re-enter the cell cycle and 60 that become post-proliferative Q cells that migrate out of the PVE. This would produce an expansion of the proliferative population by 40%. Since the VZ only increases slightly in thickness during this period, and the packing density of the cells remains constant (Takahashi et al. 1996), the bulk of this expansion results in an increase in the surface area of the VZ and, hence, in the surface area of the entire developing neocortex. As Q increases to 0.5 and P decreases to 0.5, a point is reached where "steady-state growth" is achieved transiently. At this time, the number of P cells produced is exactly enough to replenish the proliferative population; an equivalent number of Q cells are produced that leave the PVE. As development proceeds, however, a second period, when Q > 0.5 and P < 0.5, is reached. During this second period, the prolifer-

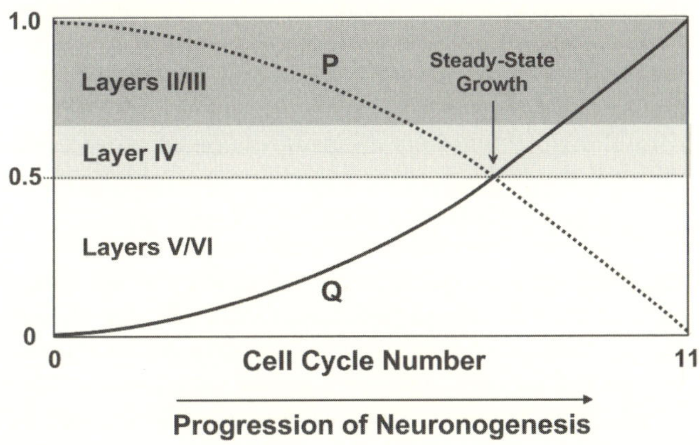

Fig. 9. Q and P are complementary (i.e., P = 1–Q), and, as Q increases from 0 to 1, P decreases from 1 to 0. When P is > 0.5, more daughter cells re-enter the cell cycle than leave it, and the neopallium will expand. In the mouse, between cell cycle 7 and 8, a point is reached when this expansion stops, and both P and Q are equal to 0.5. This occurs approximately as the cells that will reside in the vicinity of the border between layers V and IV are produced

ative population contracts. This is because fewer daughter cells re-enter the S-phase than are needed to maintain it. For example, if $Q = 0.7$ and $P = 0.3$, then for every 100 proliferating cells present at the beginning of the cell cycle only 60 will re-enter it, and 140 cells will become Q cells. This represents a 40% reduction of cell number in the PVE. Since the ventricular surface does not contract during the developmental period and the packing density of the cells remains constant (Takahashi et al. 1996), most of this must be reflected in a reduction of the thickness of the VZ. This, in fact, correlates with what is known about the development and involution of the VZ (Caviness et al. 2000; Takahashi et al. 2001).

The crossover point corresponding to the time when $P = Q = 0.5$, at which time the neocortical primordium ceases to expand, occurs as the NI passes through cell cycle 7. This point is important because it is when the expansion of the PVE stops. A cycle-by-cycle analysis of the laminar position of the neurons generated at each of the 11 cell cycles shows that this crossover point occurs as the last neurons of layer V are being produced (Fig. 10). This means that virtually all of the deep layers of the neocortex are produced during the first, expansion, phase of the NI, and that virtually all of the superficial layers are produced during the second, extinction, phase.

The pathway of changes in Q and P from $Q = 0 \rightarrow Q = 1$ and $P = 1 \rightarrow P = 0$ determines three properties of the proliferative population: (1) the lifespan of the PVE population, (2) the expansion of the PVE population and, hence, of the neocortical primordium, and (3) the output from the population, i.e., the number of neurons produced both per cycle and also during the total NI. Each of these properties can be approached quantitatively.

The lifespan of the NI is most closely regulated by changes in Q, which, as graphed above, can be described by the simple formula seen in Eq. (1), in which CC is the cell cycle, k is a dimensionless constant and a is a constant reflecting the deviation of the $Q = 0 \rightarrow Q = 1$ curve from linear over the course of the NI. For the developing mouse neocortex, k has been determined empirically to be approximately 0.009, and a is approximately 1.97. Thus, Eq. (1) describes a best-fit relationship between the progression of Q and cell cycle number (Fig. 8); however, the biology represented by these changes in Q may carry a substantially deeper significance than merely description. This is because the lifespan of the NI, i.e., the number of cell cycles that constitutes the neuronogenetic sequence, may be specified by the molecular biological mechanisms that are encapsulated in this equation. For example, developmental control of k could change the size of the "steps" of the changes in Q with each pass through the cell cycle. A smaller value of k would lengthen the NI in terms of the number of cell cycles, whereas a larger value of k will shorten it. Thus, the mechanisms which determine the changes in Q, i.e., the changes in the probability at each successive cycle that postmitotic daughter cells will exit the proliferative population, determine the number of cycles in the NI.

$$Q = k^{*}CC^{a} \qquad (1)$$

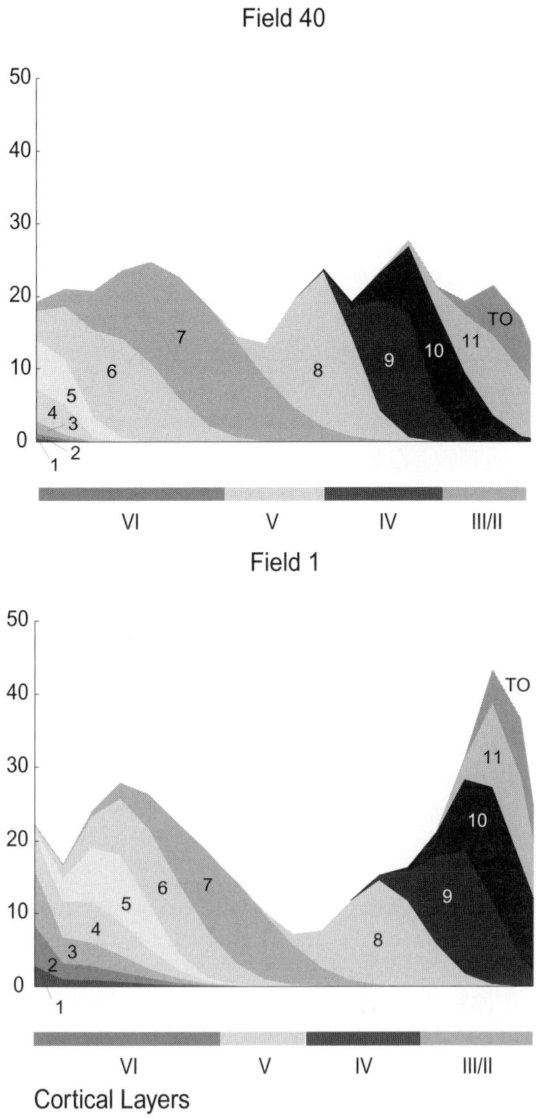

Fig. 10. Laminar distribution of neurons produced during each of the 11 cell cycles of the neuronogenetic period for two non-adjacent cytoarchitectonic areas: *field 1*, which is located dorsomedially, and *field 40*, which is located rostrolaterally. Each cell cycle produces neurons that are distributed in several layers, but there is a systematic change in the laminae of residence with each sequential cell cycle. Also, each layer receives neurons that are produced during more than one cell cycle. Note that the neurons of layers *VI* and *V* are produced during the first seven cells cycles, i.e., during the period when the neopallium is still expanding (Fig. 9)

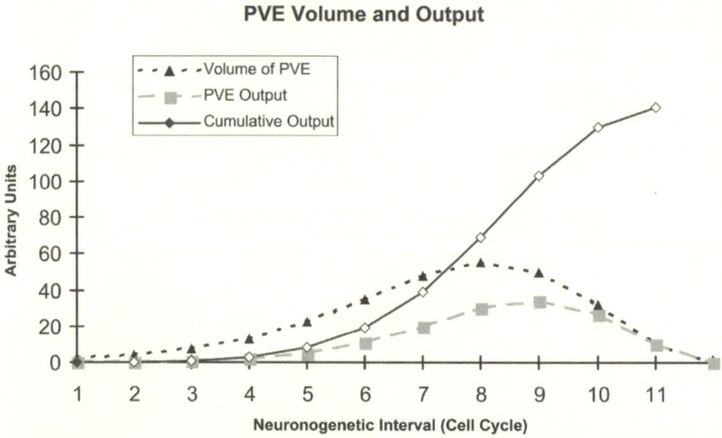

Fig. 11. Graph of the changes in the PVE volume, PVE output per cell cycle and the cumulative PVE output as given by Eqs. (2), (3), and (4) in the text. Note that the volume increases through cell cycle 8 when P first falls below 0.5 (Fig. 9)

The expansion of the PVE is also specified by the changes in Q and P. In this case, the expansion at each cell cycle is dictated by P, i.e., the probability that the daughter cells will re-enter the cell cycle. At the start of the NI, P is close to 1, so the expansion during the first few cell cycles is close to the maximum attainable, which is a doubling in cell number for each cell cycle. The actual expansion is twice the value of P at each cell cycle, and, in addition, the expansion is multiplicative at each cell cycle. For example, for the first three cell cycles of the NI, the value P (from Eq. 1) starts at 0.991 and then decreases first to 0.965, and then to 0.923. This would produce expansions of 1.982, 1.93, and 1.846, respectively. Thus, the total expansion for a "unit volume" of the PVE during these first three cell cycles is the product of these three numbers (1.982 × 1.93 × 1.846), or 7.061. In other words, during the first three cell cycles of the NI (which occur in just over 1 day), the PVE expands over sevenfold. Equation (2) shows this expansion series for the entire 11 cell cycles that comprise the NI. A graph of this series (Fig. 11) shows that it is expected that the PVE would expand over 30-fold by the end of cell cycle 7. Of course, during cell cycle 7, the "crossover" from P = Q = 0.5 occurs so that expansion stops and the PVE volume regresses to 0 at the end of the NI.

$$PVE_{cc} = P_1^{*} \prod_{i=2}^{cc} 2^{*} P_i \qquad (2)$$

The output (or the number of neurons formed) from a "single unit" of the PVE at each cell cycle is equal to twice the number of cells present in the PVE at the beginning of the cycle times Q (Eq. 3). A graph of this series (Fig. 11) shows that the predicted output rises gradually to a peak and then falls. This is, of course, expected because of the change in the size of the PVE. It is also reassuring because it matches qualitatively the shape of the curves usually obtained from counts of the percent of neurons born on sequential days during development (Angevine and Sidman 1961; Caviness and Sidman 1973; Rakic 1974; Nowakowski and Rakic 1975; Stanfield and Cowan 1979). The cumulative total output at each cell cycle is simply the sum of the output for all preceding cell cycles (Eq. 4). A graph of the cumulative output (Fig. 11) rises steadily to a total of over 140, indicating that on average a single PVE cell present at the beginning of the NI produces over 140 neurons.

$$OUT_n = 2^* PVE_n^{\ *} Q_n \qquad (3)$$

$$TOTAL_{cc} = \sum_{n=0}^{cc} OUT_n \qquad (4)$$

Both the expansion and the output of the PVE are markedly influenced by the precise pathway that is taken from P = 1→P = 0. This is because the expansion is the product of each P for each cell cycle (Eq. 2). Thus, if P descends either more quickly or more slowly from 1 than indicated by Eq. (1), both expansion and output will be affected. To relate this to Eq. (1), the constant in the exponent, "a", is the most influential on the pathway from P = 1→P = 0. If a is smaller than the value in mouse (where a = 1.97), Q will rise towards 0.5 faster and, hence, P will fall towards 0.5 faster. As a result, the PVE will expand less, and the surface area of the neocortex will be decreased; this will reduce the total number of neurons in the neocortex. Conversely, if a is greater than 1.97, Q will remain closer to 0 for more cell cycles, and P will remain closer to 1 for more cell cycles. This will produce a greater expansion of the surface area of the PVE and more neurons in the final product, i.e., in the neocortex. Thus, the two parameters in Eq. (1), k and a, can both affect the size of the neocortex and the final number of neurons independently. The combined effects of changes in these two parameters are shown in Table 2. The expansion in a

Table 2. Calculated expansion from an average founder cell, i.e., a proliferating cell present at the beginning of the NI as a function of two parameters, a (in columns) and k (in rows), from Eq. (1) that may influence the pathway from P = 1→P = 0. The shaded row and column indicate the values of a (1.97) and k (0.009) that are found in the mouse

k (cc) a	0.022 (7)	0.013 (10)	**0.009 (11)**	0.0048 (12)	0.0019 (24)	0.0014 (28)	0.00091 (35)
1.64	17.8	58.5	**85**	**123**	7,380	27,300	261,000
1.77	19.3	67.3	**99.6**	146	10,850	43,050	465,000
1.87	20.5	74.2	**111**	165	14.280	59,490	700,000
1.97	**21.6**	**81.3**	**123**	**186**	**18,400**	**80,500**	**1,030,000**
2.07	22.7	88.5	**136**	207	23,450	106,900	1,470,000
2.17	23.7	95.7	**148**	228	29,370	139,500	2,066,000
2.55	27.2	**123**	**198**	315	61,600	336,000	6,300,000

mouse is about 123-fold per founder cell which occurs during the 11th cell cycle of the NI. If k is changed so as to change the number of cell cycles of the NI without changing a (the exponent in Eq. 1), then the expansion will be affected (row of boldface values in Table 2). A slight change in k that is sufficient to increase or decrease the number of cell cycles by 1 would produce a large change in the number of neurons generated. This means that changes in the length of the cell cycle that are only about 10% would dramatically change the number of neurons produced unless compensatory changes in the length of the NI and/or in a also occurred. Changes in k that change the cell cycle numbers by greater amounts have dramatic effects on the numbers of neurons produced. For example, a 10-fold decrease in k, which is enough to lengthen the NI from 11 to 35 cell cycles, i.e., the estimated number of cell cycles needed to make the human brain, would increase the number of neurons produced by almost 10,000-fold (see Table 2)! Indeed, this amount of expansion is approximately consistent with the greater number of cells in the larger human neocortex (Caviness et al. 1995).

The changes in the size of the PVE and developing cortical plate and the relationships of these changes can be seen more easily if the equations and line graphs shown above are converted into a schematic form (Fig. 12). Figure 12

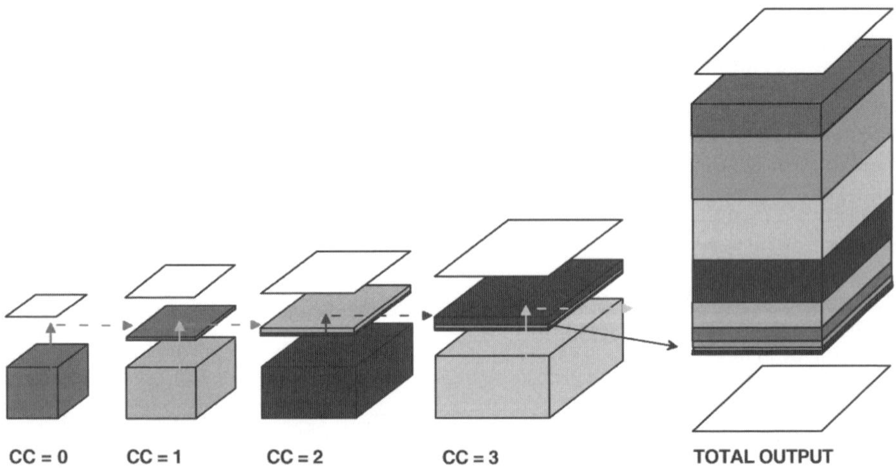

Fig. 12. Visualization of the changes shown in the graphs of Fig. 11 and as given by Eqs. (2), (3), and (4) in the text. At the onset of the neuronogenetic interval (CC = 0), a single unit of the PVE is shown. At the next cell cycle (CC = 1), the PVE has an increased volume; the output from the first cell cycle is shown in the position of the cortical plate. At CC = 2, the PVE has increased in volume again, and now the output from the first two cell cycles is shown in the position of the cortical plate. At CC = 3, the process is repeated. On the *right-hand side* of the figure, the diagram shows the final total output of all of the 11 cell cycles of the neuronogenetic interval. Note that the output from the first three cell cycles corresponds to only a small part of the total output, whereas the output of the last three cell cycles comprises about 50% of the total output

shows the cycle-by-cycle growth of the PVE and the cortical plate for the first three cell cycles of the NI. At the beginning of the NI, the only cells present are the cells of the PVE; they are represented as a cube with a "unit volume" that is 1 unit high, 1 unit wide and 1 unit deep. Since the PVE is about 6 cells high, such a unit volume would contain about 6^3 or 216 cells. During the first cell cycle P is about 0.99 and Q is about 0.01. Thus, most of the daughter cells produced will remain in the proliferative population, and the PVE will expand. From histological observations (Takahashi et al. 1996), we know that the VZ does not increase much in thickness during the NI; therefore, most of the expansion of the unit of the PVE will be in the dimensions parallel to the surface of the lateral ventricle. Thus, at the beginning of the second cell cycle, the unit of PVE will be 1.98 times its volume and will have produce only 0.02 units. For convenience, we illustrate the output in the position of the future cortical plate as if it had completed its migration instantaneously. At each of the next two cell cycles, P remains high, and most of the daughter cells remain in the PVE, but each cell cycle contributes an increasing amount to the developing cortical plate. At the end of the third cell cycle, the unit of the PVE has expanded to over seven times its original volume, and produced about 1% of the neurons that will comprise the neocortex in the adult. Most of the cells of the neocortex are produced during the last few cell cycles (Fig. 12).

The neocortex VZ contains many of these "units" arrayed across its surface, and both the sequence of 11 cell cycles and the changes in P and Q in each are identical, at least to the resolution that has been used so far (Miyama et al. 1997). The result of this arrangement is that different events in the sequence occur contemporaneously in different regions of the VZ (Fig. 13). The NI is initiated first in the rostrolateral cortex, and, given the fact that there is more than a 24-h difference between the rostrolateral cortex and the dorsomedial cortex (Miyama et al. 1997), when the NI is first initiated in the dorsomedial cortex, the rostrolateral cortex has already progressed into cell cycle 3 or even 4. This means that there is a gradient of maturation beginning in the rostrolateral cortex and spreading across the surface of the developing cortex. The gradient of maturation means that Tc, T_{G1}, Q and P differ across the surface of the developing cortex (Fig. 13). Thus, at any given time, the status of these proliferatively related parameters provides positional information. From the perspective of the cell cycle, this gradient divides the surface of the developing PVE into "cell cycle domains", i.e., regions of the PVE in which all of the PVE cells are in the same cell cycle. As the developing cortex matures and each "unit" of the PVE progresses through the NI, these cell cycle domains "move" across the surface of the PVE defining a series of "waves" that radiate from the striatocortical fissure at the lateral edge of the neopallium (Fig. 13). We speculate that the positional information in these cell cycle domains may, in coordination with an unknown synchronizing signal, be sufficient to define the cytoarchitectonic borders that characterize the adult neocortex (Brodmann 1909; Caviness 1975). Two possible candidates that have properties appropriate to serve as a synchronizing signal are electrotonic coupling of the cells of the PVE (Bittman et

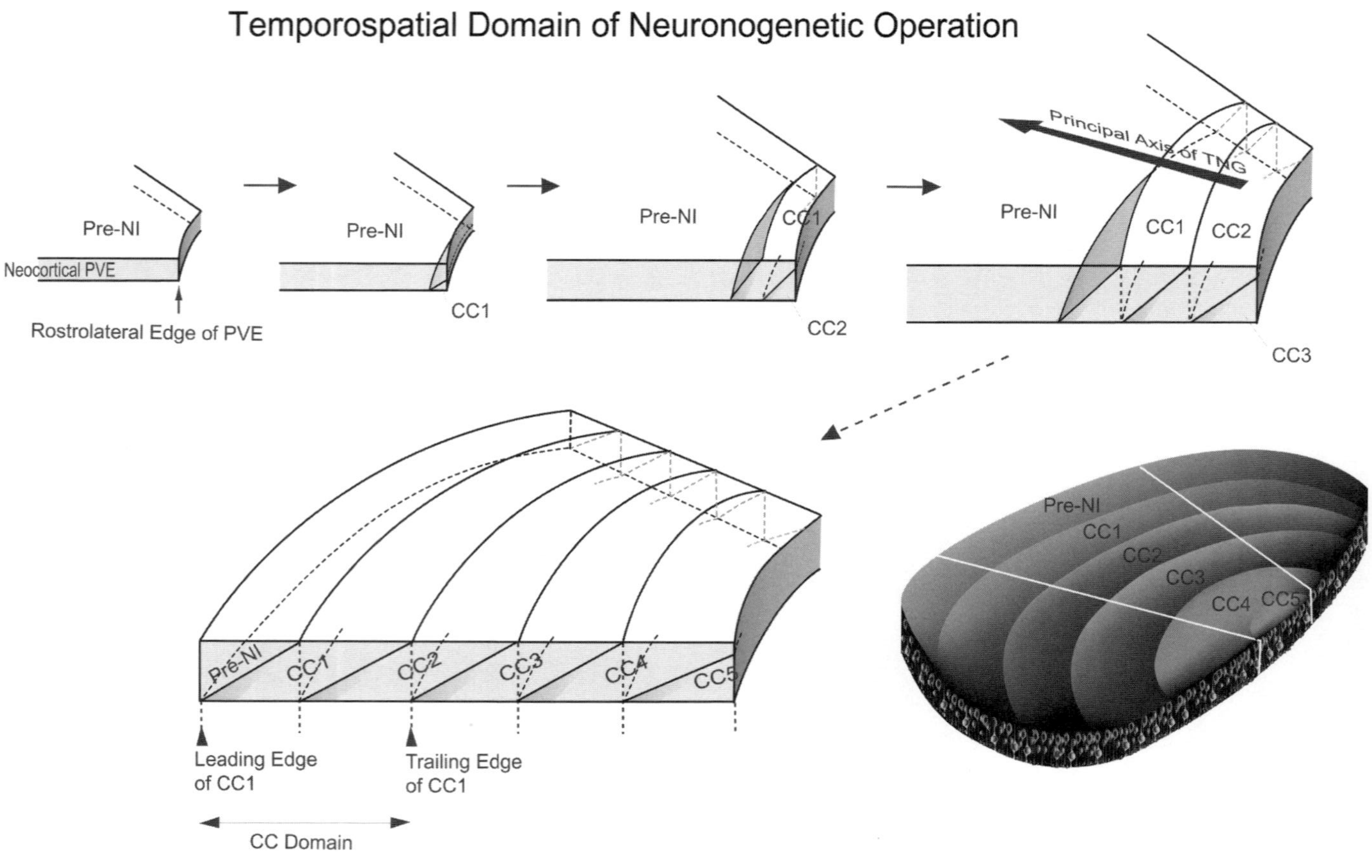

Fig. 13. Sequence of dynamic changes in the length of the cell cycle (and in P and Q) is initiated in the rostrolateral-most portions of the neopallium and then spreads as a gradient of maturation across the neopallial surface. This wave-like progression of maturation means that, at any given time, there are "domains" of the PVE that are in different states. This is, in theory, sufficient to provide a basis for cell cycle length to serve as positional information that could be involved in the development of cytoarchitectonic subdivisions. For details, see the text

al. 1997) or perhaps afferents into the neopallium. We envision that the synchronizing signal and the "waves" in the PVE interact to influence the fate of the neurons generated at the time that the interactions occur. Since neocortical mapping seems to be dependent on information that resides in the deeper layers (V and VI; De Carlos and O'Leary 1992; Erzurumlu and Jhaveri 1992; Molnar and Blakemore 1995), and since the cells destined for these layers are produced during the first half of the neuronogenetic period (Figs. 9 and 10), it would seem to be this period when the positional information has its greatest effect. The conversion of the positional information present as a difference in the length of G1 into directives controlling cell fate could be carried out by creating differences in the transcriptional profiles of the cells. There is evidence that short cell cycles can prevent the completion of transcription of some genes (Rothe et al. 1992) and that some large genes involved in CNS function take up to 16 h to be transcribed (Tennyson et al. 1995). Taken together, this provides a simple and elegant way for cell cycle changes in the PVE to be "mapped" onto the post-proliferative population.

The intercellular, intracellular and molecular controls on the dynamic changes in the proliferating cells of the PVE are, at present, poorly understood. It is likely, however, that they are complex and changing, i.e., that different influences predominate at different stages of the NI. A tantalizing hint into this complexity is provided by an analysis of the time and concentration of the mRNA of two cell cycle associated molecules, cyclin E and $p27^{Kip1}$ (Fig. 14; Delalle et al. 1999). As expected, at the onset of the NI, the level of cyclin E

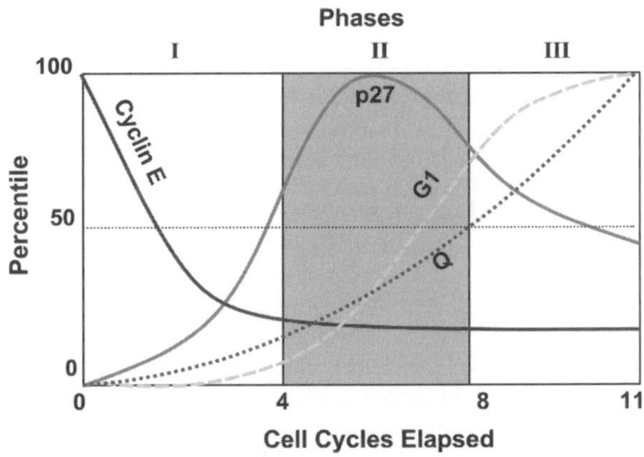

Fig. 14. Schematic graphs of the changes in the levels of expression of cyclin E and p27 along with the changes in the length of G1 and in Q that occur during the 11 cell cycles of the neuronogenetic interval (NI). The levels of cyclin E and p27 change during the NI and define three phases during which cyclin E is high and p27 is low, cyclin E is low and p27 is high, and when both are low. We suggest that other molecular regulators of the cell cycle are active during these phases, especially during the third phase, to effect the systematic changes in the length of G1 and in Q and P

starts out high; during the first three to four cell cycles, however, it falls. At the same time, $p27^{Kip1}$ levels, which start low, rise and continue to rise to a peak near the middle of the NI. After this, $p27^{Kip1}$ levels fall. The rise and fall of these two mRNAs divides the NI into three phases: an early phase when cyclin E is high and $p27^{Kip1}$ is low, a middle phase when cyclin E is low and $p27^{Kip1}$ is high, and a late phase when both cyclin E and $p27^{Kip1}$ expression levels are low (Fig. 14). Note that the changes in the levels of cyclin E and $p27^{Kip1}$ are not related in any simple way to either changes in Q or changes in T_{G1}. The role of $p27^{Kip1}$ in lengthening G1 has also recently been established using an inducible promoter system (Mitsuhashi et al. 2001). Overexpression of $p27^{Kip1}$ during the middle phase of the NI leads to a lengthening of the G1-phase of cortical PVE cells from about 9 to 13 h (Mitsuhashi et al. 2001). The continued rise in G1 and Q despite the fall in $p27^{Kip1}$ indicates that other molecules must come into play during the third phase. This means that other cell cycle regulators are probably involved (Zindy et al. 1999; Cunningham and Roussel 2001). In addition, it has been shown that growth factors and other small molecules, including FGF2 (Ghosh and Greenberg 1995; Vaccarino et al. 1999a,b; Raballo et al. 2000), PACAP (Nicot and DiCicco-Bloom 2001; Suh et al. 2001), IGF-1 (Drago et al. 1991), and NT3 (Ghosh and Greenberg 1995), may all play a role either as a mitogen or an anti-mitogen. In addition, proliferating cells in the ventricular zone are interconnected by gap junctions (Bittman et al. 1997; Bittman and LoTurco 1999) and express GABA(A) receptors indicating (Owens et al. 1999) that cell-cell signaling (Owens et al. 2000) may also play a role in regulating the complex and dynamic behaviors of the proliferating cells of the ventricular zone.

References

Angevine JBJ, Sidman RL (1961) Autoradiographic study of cell migration during histogenesis of cerebral cortex in the mouse. Nature 192:766–768

Beaulieu C (1993) Numerical data on neocortical neurons in adult rat, with special reference to the GABA population. Brain Res 609:284–292

Beaulieu C, Colonnier M (1989) Number of neurons in individual laminae of areas 3B, 4 gamma, and 6a alpha of the cat cerebral cortex: a comparison with major visual areas. J Comp Neurol 279:228–234

Bittman KS, LoTurco JJ (1999) Differential regulation of connexin 26 and 43 in murine neocortical precursors. Cereb Cortex 9:188–195

Bittman KS, Owens DF, Kriegstein AR, LoTurco JJ (1997) Cell coupling and uncoupling in the ventricular zone of developing neocortex. J Neurosci 17:7037–7044

Blau HM, Brazelton TR, Weimann JM (2001) The evolving concept of a stem cell: entity or function? Cell 105:829–841

Boulder Committee (1970) Embryonic vertebrate central nervous system: revised terminology. Anat Rec 166:257–262

Brodmann K (1909) Vergleichende Lokalisationslehre der Grosshirnrinde. Barth, Leipzig

Cai L, Hayes NL, Nowakowski RS (1997) Local homogeneity of cell cycle length in developing mouse cortex. J Neurosci 17:2079–2087

Caviness VS, Sidman RL (1973) Time of origin of corresponding cell classes in the cerebral cortex of normal and reeler mutant mice: an autoradiographic analysis. J Comp Neurol 148:141–152

Caviness VS, Takahashi T, Nowakowski R (1995) Numbers, time and neocortical neuronogenesis: a general developmental and evolutionary model. Trends Neurosci 18:379–383

Caviness VS Jr, Takahashi T, Nowakowski RS (2000) Neuronogenesis and the early events of neocortical histogenesis. Res Prob Cell Differ 30:107–143

Caviness VSJ (1975) Architectonic map of neocortex of the normal mouse. J Comp Neurol 164: 247–264

Chenn A, McConnell S (1995) Cleavage orientation and the asymmetric inheritance of Notch1 immunoreactivity in mammalian neurogenesis. Cell 82:631–641

Chenn A, Zhang YA, Chang BT, McConnell SK (1998) Intrinsic polarity of mammalian neuroepithelial cells. Mol Cell Neurosci 11:183–193

Cunningham JJ, Roussel MF (2001) Cyclin-dependent kinase inhibitors in the development of the central nervous system. Cell Growth Differ 12:387–396

De Carlos JA, O'Leary DD (1992) Growth and targeting of subplate axons and establishment of major cortical pathways. J Neurosci 12:1194–1211

Delalle I, Takahashi T, Nowakowski RS, Tsai LH, Caviness VS Jr (1999) Cyclin E-p27 opposition and regulation of the G1 phase of the cell cycle in the murine neocortical PVE: a quantitative analysis of mRNA in situ hybridization. Cereb Cortex 9:824–832

Drago J, Murphy M, Carroll SM, Harvey RP, Bartlett PF (1991) Fibroblast growth factor-mediated proliferation of central nervous system precursors depends on endogenous production of insulin-like growth factor I. Proc Natl Acad Sci USA 88:2199–2203

Erzurumlu RS, Jhaveri S (1992) Emergence of connectivity in the embryonic rat parietal cortex. Cereb Cortex 2:336–352

Ghosh A, Greenberg ME (1995) Distinct roles for bFGF and NT-3 in the regulation of cortical neurogenesis. Neuron 15:89–103

Gratzner HG (1982) Monoclonal antibody to 5-bromo- and 5-iododeoxyuridine: a new reagent for detection of DNA replication. Science 218:474–478

Hamilton E, Dobbin J (1983a) The percentage labeled mitosis technique shows the mean cell cycle time to be half its true value in carcinoma TY. I. [H3]Thymidine and vincristine studies. Cell Tissue Kinet 16:473–482

Hamilton E, Dobbin J (1983b) The percentage labeled mitosis technique shows the mean cell cycle time to be half its true value in carcinoma NT. II. [3H]Deoxyuridine studies. Cell Tissue Kinet 16:483–492

Haug H (1987) Brain sizes, surfaces, and neuronal sizes of the cortex cerebri: a stereological investigation of man and his variability and a comparison with some mammals (primates, whales, marsupials, insectivores, and one elephant). Am J Anat 180:126–142

Hayes NL, Nowakowski RS (2000) Exploiting the dynamics of S-phase tracers in developing brain: interkinetic nuclear migration for cells entering versus leaving the S-phase. Dev Neurosci 22: 44–55

Hoshino K, Matsuzawa T, Murakami U (1973) Characteristic of the cell cycle of matrix cells in the mouse embryo during histogenesis of telencephalon. Exp Cell Res 77:89–94

Kaufmann SL (1968) Lengthening of the generation cycle during embryonic differentiation of the mouse neural tube. Exp Cell Res 49:420–424

Miller MW, Nowakowski RS (1988) Use of bromodeoxyuridine-immunohistochemistry to examine the proliferation, migration and time of origin of cells in the central nervous system. Brain Res 457:44–52

Mitsuhashi T, Aoki Y, Eksioglu YZ, Takahashi T, Bhide PG, Reeves SA, Caviness VS Jr (2001) Overexpression of p27Kip1 lengthens the G1 phase in a mouse model that targets inducible gene expression to central nervous system progenitor cells. Proc Natl Acad Sci USA 98:6435–6440

Miyama S, Takahashi T, Nowakowski RS, Caviness VS Jr (1997) A gradient in the duration of the G1 phase in the murine neocortical proliferative epithelium. Cereb Cortex 7:678–689

Molnar Z, Blakemore C (1995) How do thalamic axons find their way to the cortex? Trends Neurosci 18:389–397

Nicot A, DiCicco-Bloom E (2001) Regulation of neuroblast mitosis is determined by PACAP receptor isoform expression. Proc Natl Acad Sci USA 98:4758–4763

Nowakowski RS, Rakic P (1974) Clearance rate of exogenous 3H-thymidine from the plasma of pregnant rhesus monkeys. Cell Tissue Kinet 7:189–194
Nowakowski RS, Rakic P (1975) Time of origin of neurons in the hippocampal region of the rhesus monkey. Neurosci Abstr 1:773
Nowakowski RS, Lewin SB, Miller MW (1989) Bromodeoxyuridine immunohistochemical determination of the lengths of the cell cycle and the DNA-synthetic phase for an anatomically defined population. J Neurocytol 18:311–318
Owens DF, Liu X, Kriegstein AR (1999) Changing properties of GABA(A) receptor-mediated signaling during early neocortical development. J Neurophysiol 82:570–583
Owens DF, Flint AC, Dammerman RS, Kriegstein AR (2000) Calcium dynamics of neocortical ventricular zone cells. Dev Neurosci 22:25–33
Raballo R, Rhee J, Lyn-Cook R, Leckman JF, Schwartz ML, Vaccarino FM (2000) Basic fibroblast growth factor (Fgf2) is necessary for cell proliferation and neurogenesis in the developing cerebral cortex. J Neurosci 20:5012–5023
Rakic P (1974) Neurons in rhesus monkey visual cortex: systematic relation between time of origin and eventual disposition. Science 183:425–427
Rockel AJ, Hiorns RW, Powell TP (1974) Proceedings: numbers of neurons through full depth of neocortex. J Anat 118:371
Rockel AJ, Hiorns RW, Powell TP (1980) The basic uniformity in structure of the neocortex. Brain 103:221–244
Rothe M, Pehl M, Taubert H, Jackle H (1992) Loss of gene function through rapid mitotic cycles in the Drosophila embryo. Nature 359:156–159
Sauer FC (1936) The interkinetic migration of embryonic epithelial nuclei. J Morphol 60:1–11
Sidman RL (1970) Autoradiographic methods and principles for study of the nervous system with thymidine-H3. In: Nauta WJH, Ebbesson SOE (eds) Contemporary research methods in neuroanatomy. Springer, Berlin Heidelberg New York, pp 252–274
Sidman RL, Miale IL, Feder N (1959) Cell proliferation and migration in the primitive ependymal zone: an autoradiographic study of histogenesis in the nervous system. Exp Neurol 1:322–333
Stanfield BB, Cowan WM (1979) The development of the hippocampus and dentate gyrus in normal and reeler mice. J Comp Neurol 185:423–459
Suh J, Lu N, Nicot A, Tatsuno I, DiCicco-Bloom E (2001) PACAP is an anti-mitogenic signal in developing cerebral cortex. Nat Neurosci 4:123–124
Takahashi T, Nowakowski RS, Caviness VS Jr (1992) BUdR as an S-phase marker for quantitative studies of cytokinetic behaviour in the murine cerebral ventricular zone. J Neurocytol 21:185–197
Takahashi T, Nowakowski RS, Caviness VS Jr (1993) Cell cycle parameters and patterns of nuclear movement in the neocortical proliferative zone of the fetal mouse. J Neurosci 13:820–833
Takahashi T, Nowakowski RS, Caviness VS Jr (1995) The cell cycle of the pseudostratified ventricular epithelium of the embryonic murine cerebral wall. J Neurosci 15:6046–6057
Takahashi T, Nowakowski RS, Caviness VS Jr (1996) The leaving or Q fraction of the murine cerebral proliferative epithelium: a general model of neocortical neuronogenesis. J Neurosci 16:6183–6196
Takahashi T, Nowakowski RS, Caviness VS Jr (2001) Neocortical neuronogenesis: regulation, control points and a strategy of structural variation. In: Nelson CA, Luciana M (eds) Handbook of cognitive developmental neuroscience. MIT Press, Cambridge, MA, pp 3–22
Tennyson CN, Klamut HJ, Worton RG (1995) The human dystrophin gene requires 16 hours to be transcribed and is cotranscriptionally spliced. Nat Genet 9:184–190
Vaccarino FM, Schwartz ML, Raballo R, Rhee J, Lyn-Cook R (1999a) Fibroblast growth factor signaling regulates growth and morphogenesis at multiple steps during brain development. Curr Top Dev Biol 46:179–200

Vaccarino FM, Schwartz ML, Raballo R, Nilsen J, Rhee J, Zhou M, Doetschman T, Coffin JD, Wyland JJ, Hung YT (1999b) Changes in cerebral cortex size are governed by fibroblast growth factor during embryogenesis. Nat Neurosci 2:848

Zindy F, Cunningham JJ, Sherr CJ, Jogal S, Smeyne RJ, Roussel MF (1999) Postnatal neuronal proliferation in mice lacking Ink4d and Kip1 inhibitors of cyclin-dependent kinases. Proc Natl Acad Sci USA 96:13462–13467

Mechanisms Regulating Lineage Diversity During Mammalian Cerebral Cortical Neurogenesis and Gliogenesis

Mark F. Mehler[1]

Summary. During mammalian cerebral cortical development, neural stem cells (NSCs) present within periventricular generative zones give rise to successive waves of neurons and radial glia, followed by oligodendrocytes and astrocytes. The molecular and cellular mechanisms that orchestrate these precisely timed and progressive maturational events are still largely undefined. These developmental processes are likely to involve the dynamic interplay of environmental signals, cell-cell interactions and transcriptional regulatory events. The bone morphogenetic proteins (BMPs), an expanding subclass of the transforming growth factor β cytokine superfamily, may represent an important set of environmental cues for these progressive maturational events because of the broad profiles of developmental expression of the requisite BMP ligands, receptor subunits and intracellular transduction elements, and because of their versatile roles in promoting a spectrum of cellular processes intimately involved in progressive neural fate decisions. The BMPs also interact with complementary regional environmental signals such as the basic fibroblast growth factor (bFGF) and sonic hedgehog (Shh) that promote earlier stages of NSC expansion, self-renewal, lineage restriction and incipient lineage commitment. The ability of these cytokines and trophic signals to act within specific neurodevelopmental contexts may, in turn, depend on the composite actions of cell-cell contact-associated signals, such as Notch-Hes-mediated lateral inhibitory pathways, and additional transcriptional modulatory events, such as those mediated by members of the inhibitor of differentiation (ID) gene family that encode a novel set of negative basic helix-loop-helix (bHLH) transcription factors. In this chapter, we will examine the distinct roles of these different classes of developmental cues in defining the biological properties of an integrated cerebral cortical developmental signaling network. Ongoing studies in this exciting area of mammalian central nervous system (CNS) development will help to identify important molecular and cellular targets for evolving pharmacological, gene and stem cell therapeutic interventions to combat the pathological sequelae of a spectrum of acquired and genetic disorders of the central nervous system.

1 Stem Cell Biology and Neural Development

The major cellular elements of the mature brain, including neurons, radial glia, oligodendrocytes (OLs) and astrocytes, arise predominantly from specialized generative zones that are derived from the inner lining of the neural tube (Gage et al. 1995; Weiss et al. 1996; McKay 1997). These zones, extending from

[1] F.M. Kirby Program in Neural Protection and Repair, Departments of Neurology, Neuroscience and Psychiatry, Rose F. Kennedy Center for Research in Mental Retardation and Developmental Disabilities, Einstein Comprehensive Cancer Center, Albert Einstein College of Medicine, 1410 Pelham Parkway South, Bronx, New York 10461, USA

Results and Problems in Cell Differentiation, Vol. 39
C.F. Hohmann (Ed.): Cortical Development

periventricular regions of the telencephalon to the spinal cord, consist of a pseudostratified epithelium, the ventricular zone (VZ), that gives rise during later embryonic life to secondary subventricular zones (SVZs) that persist in attenuated form into adult life (Morshead et al. 1994; Weiss et al. 1996a,b; McKay 1997). Neurons and radial glia are generated predominantly from early embryonic VZ progenitor cells, whereas oligodendrocytes and astrocytes are also generated from late embryonic and perinatal SVZ progenitors (Cameron and Rakic 1991; Goldman et al. 1997). Neural stem cells (NSCs) are generally defined by their ability to undergo constitutive proliferation, to exhibit self-maintenance (self-renewal), to generate large numbers of more lineage-restricted progeny through transient amplification of intermediate progenitor pools and to generate new cells in response to injury or specific disease states (Reynolds et al. 1992; Gage et al. 1995; Reynolds and Weiss 1996; Kalyani et al. 1997; McKay 1997; Morrison et al. 1997). The complement of NSCs is regulated primarily by the kinetics of proliferation (McKay 1997). This mitotic profile, in turn, is a reflection of cell cycle duration and the ratios of symmetric to asymmetric cell divisions (Caviness and Takahashi 1995; McKay 1997). Maintenance of the NSC phenotype is mediated by both daughter cells during the period of symmetric cell division and rapid expansion of the stem cell pool during early embryogenesis, and by one daughter cell of each pair during later asymmetric cell divisions (Caviness and Takahashi 1995; McKay 1997). NSC maintenance involves cell intrinsic as well as extrinsic mechanisms (Henrique et al. 1995; McKay 1997). Through a process of lateral inhibition, the Notch signaling pathway regulates the precise timing of neurogenesis and gliogenesis while maintaining a cohort of NSCs (Quian et al. 1998; Wang et al. 1998). The Notch effectors, Hes 1 and 5, negatively regulate transcriptional activation by neurogenic bHLH factors, and normally function to maintain NSC self-renewal and also to repress commitment of multipotent progenitors to the neuronal lineage during the phase of asymmetric cell division (Ishibashi et al. 1994, 1995; Ohtsuka et al. 1999; Nakamura et al. 2000).

By contrast, following activation by neural determination bHLH genes (i.e., Mash 1), Hes 6 promotes neurogenesis through inhibitory effects on Hes 1, and by positive feedback regulation of Mash 1 through Notch-independent pathways (Bae et al. 2000; Koyano-Nakagawa et al. 2000). Mammalian numb, a Notch antagonist, preserves the NSC state by preferential segregation to apical daughter cells during asymmetric cell division (Zhong et al. 2000). Numb also functions in later stages of neuronal differentiation by modulating nuclear translocation of the intracellular domain of the Notch receptor (Zhong et al. 2000). The expression of Notch ligands (i.e., Delta) on differentiating neuroblasts serves to activate Notch receptors on ventricular zone NSCs to promote the initial generation of radial glia, and exhibit cooperative interactions with the neuregulin, glial growth factor (GGF) to enhance radial glial elongation (Gaiano et al. 2000). Activation of Notch receptors causes irreversible loss of neurogenic potential, and can trigger glial differentiation through a molecular mechanism that is dominant to the actions of the bone morphogenetic proteins (BMPs; Morrison et al. 2000). Through the mediation of a distinct Notch ligand (i.e., Jagged 1) expressed on the axons of evolving neuroblast, Notch activation also inhibits the differentiation of OL progenitors, thereby regulating neuronal-glial cell-cell matching and the appropriate spatiotemporal elaboration of myelinating OLs (Wang et al. 1998; Wang and Barres 2000). Similar glial maturational inhibitory actions of Jagged 1 also expressed on differentiated OLs serves to prevent the depletion of glial precursors, thereby maintaining a reservoir of reactive glial-restricted progenitor species to allow preservation of the NSC state (Bongarzone et al. 2000). NSC fate is also maintained by members of the ID (*i*nhibitor of *d*ifferentiation) gene family (Lyden et al. 1999). These factors can differentially regulate the stoichiometry of neurogenic bHLH factors by preferentially binding to and sequestering a subset of ubiquitous bHLH proteins, thereby preventing these molecules from forming appropriate heterodimers with cell-type-specific neurogenic bHLH factors that normally promote early neurogenesis (Norton et al. 1998; Sasai 1998; Bounpheng et al. 1999; Lyden et al. 1999).

The proliferation, survival and differentiation of NSCs and nascent progenitor species are regulated by specific cytokines, including epidermal growth factor (EGF) and bFGF (Kilpatrick and Bartlett 1995; Kilpatrick et al. 1995; Burrows et al. 1997; Qian et al. 1997, 1998; Shen et al. 1998). Both bFGF and EGF simultaneously promote the continued expression of the neuroepithelial marker, nestin, and inhibit the induction of Mash 1 and Prox 1, the latter a mammalian homologue of the prospero transcriptional regulator involved in cell cycle exit and the onset of terminal neuronal differentiation (Torii et al. 1999). Cumulative experimental evidence suggests that early embryonic cerebral cortical progenitor species are multipotent and responsive to bFGF, although they appear predisposed towards neuronal and radial glial differentiation (Kilpatrick and Barlett 1995; Qian et al. 1997, 1998; Sheng et al. 1998; Gokhan et al. 2000). However, exposure of these cells to low concentrations of bFGF favors neuronal differentiation, whereas threshold concentrations favor oligodendroglial and in some cases astroglial differentiation (Qian et al. 1997). Although early embryonic VZ cells do not express the EGF receptor or respond to appropriate ligands, the EGF receptor is progressively expressed during later developmental stages by SVZ progenitor cells, and EGF and transforming growth factor-α regulate survival and proliferation within this perinatal progenitor pool (Threadgil et al. 1995; Burrows et al. 1997; Kornblum et al. 1998). EGF-responsive cells appear to be derived from FGF-responsive progenitor species, although they exhibit different population kinetics (Vescovi et al. 1993; Ciccolini and Svendsen 1998; Martens et al. 2000). EGF-responsive progenitors are the predominant species present during perinatal gliogenesis; they retain the ability to generate all cellular phenotypes although they appear predisposed toward glial differentiation (Lazar and Blum 1992; Ferrer et al. 1996; Kornblum et al. 1997, 1998).

2
Neural Lineage Elaboration and Bone Morphogenetic Proteins

The generation of differentiated progeny from mammalian NSCs is governed by at least three distinct cellular mechanisms, each under the influence of specific growth factor signals (Ahmed et al. 1995; Gross et al. 1996; Johe et al. 1996;Shah et al. 1994, 1996; Morrison et al. 1997; Marmur et al. 1998a; Mabie et al. 1999). NSCs can generate individual lineages by probabilistic (stochastic) mechanisms, and growth factors can function to potentiate the survival and/or proliferation of cells committed to a specific lineage (selective process). Alternatively, growth factors can preferentially promote the maturation of progenitor species already committed to a specific lineage without altering cellular survival or mitogenic potential. Finally, certain growth factors can enhance the commitment of NSCs to a single lineage at the expense of alternate lineages (instructive process). We have shown that bone morphogenetic proteins (BMPs) promote the generation of neurons from early embryonic cerebral cortical VZ progenitors and the generation of astrocytes from late embryonic SVZ progenitors by an instructive mechanism (Fig. 1; Gokhan et al. 2000; Mehler et al. 2000). These findings support an evolving theme in developmental neurobiology, namely that progenitor cell responsiveness to environmental cues changes during neural development (Zhu et al. 1999a; Mehler et al. 2000; Molne et al. 2000). Our observations further suggest that these changing BMP-mediated actions on individual neural lineages involve not only positive promotion of specific pathways of differentiation, but also active suppression of alternate lineage fates (Gokhan et al. 2000).

BMPs encompass at least five major factor subgroups, and represent the most rapidly expanding subclass of the transforming growth factor-β superfamily (Mehler et al. 1997). The BMPs, multifunctional regulators of central and peripheral nervous system development, are essential for orchestrating neural induction, dorsoventral patterning of the neural tube, regional developmental apoptosis, neurogenesis and gliogenesis, cellular migration, neuronal subtype specification and terminal differentiation (Graham et al. 1996; Gross et al. 1996; Hogan 1996; Shah et al. 1996; Furuta et al. 1997; Mehler et al. 1997; Song et al. 1998; Iantosca et al. 1999; Mabie et al. 1997, 1999; Shou et al. 1999). BMPs impart developmental signals by autocrine, paracrine and secondary inductive actions, and through their roles as gradient morphogens (Hogan 1996; Nguyen et al. 2000). Complex morphogenetic gradients are orchestrated, in part, by the actions of several classes of BMP-binding proteins, including noggin, chordin, follistatin and members of the Dan family, that antagonize BMP signaling by exhibiting differential ligand binding affinities (Mehler et al. 1997; Massague et al. 2000). BMPs exert their biological effects by binding predominantly to type I (BMPRIA, BMPRIB) and type II (BMPRII) receptor subunits and by transducing intracellular signals through BMP subclass-selective receptor-

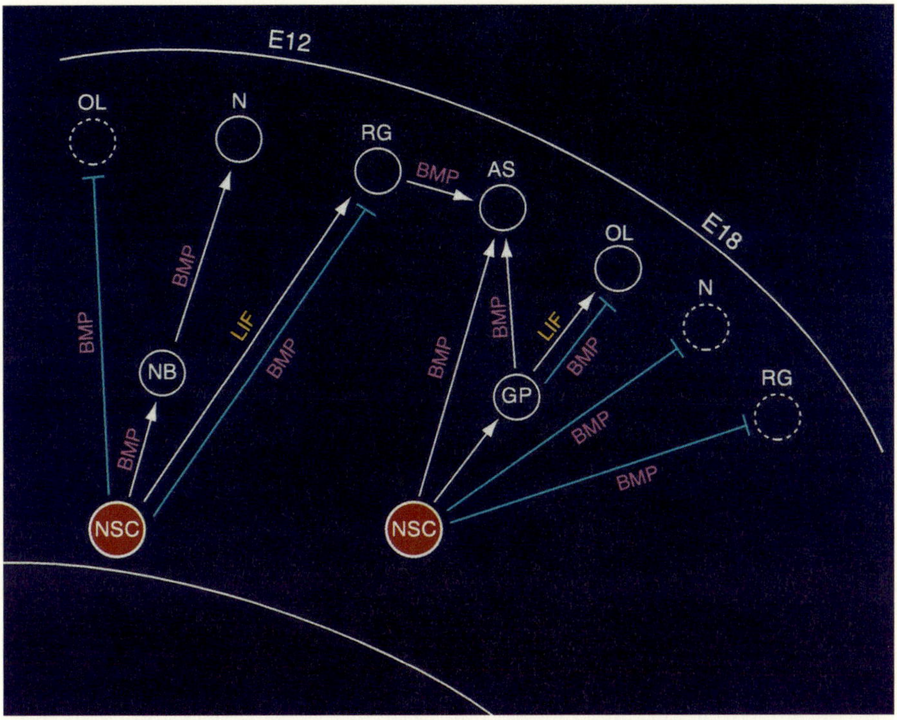

Fig. 1. Developmental model of the environmental regulation of mammalian cerebral cortical neurogenesis and gliogenesis. During the period of neurogenesis (murine, embryonic day, *E12*), neural stem cells (*NSC*) present within the early embryonic ventricular zone normally give rise to neurons (*N*) and radial glia (*RG*, both *solid circles*) but not oligodendrocytes (*OL*) or astrocytes (*AS*, both *interrupted circles*). By contrast, during the stage of late embryonic gliogenesis (*E18*), NSCs present in secondary subventricular zones give rise to astrocytes and oligodendrocytes, and cease production of neurons and radial glia. Astrocytes may be elaborated from multiple cellular precursor populations, including NSCs, radial glia and glial-restricted progenitors (*GP*). Bone morphogenetic proteins (*BMP*) exhibit both positive promotion (*white arrows*) and concurrent active suppression (*blue bars*) of specific neural lineages that depend on the developmental stage of NSCs and of their more lineage-restricted progeny. For example, during early embryonic neurogenesis, BMPs enhance the elaboration of neuroblasts (*NB*) and neurons, and inhibit the generation of radial glia and OLs. By contrast, during the phase of late embryonic gliogenesis, BMPs potentiate the generation of astrocytes from NSCs, radial glia and glial-restricted progenitors, and impede the elaboration of OLs, neurons and radial glia. The profiles of neural lineage elaboration during each seminal developmental transition represent, in part, the composite actions of multiple environmental signaling pathways. Thus, the elaboration of radial glia during the peak phase of cerebral cortical neurogenesis and the generation of OLs during the phase of gliogenesis represent a modification of developmental stage-specific BMP inhibitory signals by leukemia inhibitory factor (*LIF*). By contrast, the presence of unopposed BMP inhibitory signals effectively precludes the inappropriate elaboration of neuronal and glial species at specific developmental phases during which their presence would result in serious morphogenetic alterations and cause dysfunction of specific evolving neural networks. These developmental signaling events and lineage interrelationships are embedded within a complex network of additional complementary, cooperative, sequential and antagonistic cell intrinsic and environmental regulatory cues whose concerted actions are required to progressively sculpt the evolution of neural lineage species from early and late embryonic NSC populations (see text for details)

activated latent transcription factors (rSmads) that are themselves subject to complex transcriptional regulatory control (Soderstrom et al. 1996; Ebendal et al. 1998; Massague 1998; Whitman 1998; Kawabata and Miyazono 1999; Wrana 2000). Thus, BMP-mediated signaling pathways impart context-dependent developmental effects and exhibit combinatorial profiles of actions by employing multi-level cellular modulatory mechanisms and feedback loops that encompass the extracellular ligands, plasmalemmal receptor complexes, cytoplasmic transduction elements, additional interacting cytokine signaling pathways and the transcriptional apparatus of diverse target genes (Massague et al. 2000). Cell- and developmental-stage-specific factors determine the selection of Smad target genes, and the activation state of the evolving Smad complexes are thereafter dictated by the profiles of recruitment of distinct classes of transcriptional co-activators and co-repressors (Massague et al. 2000).

We and others have demonstrated that BMP ligands and BMP receptor subunits are expressed at the appropriate developmental times and in the appropriate CNS regions and cell types to mediate their diverse range of developmental actions (Mehler and Kessler 1996; Soderstrom et al. 1996; Mehler et al. 1997; Ebendal et al. 1998; Zhang et al. 1998). For example, BMP type I and type II receptor subunits are expressed within the early embryonic cerebral cortical VZ and within the late embryonic SVZ, as well as in regions of the evolving cortical plate (Zhang et al. 1998). Maximal expression of all BMP receptor subunits occurs during the period of late embryonic and perinatal stages of corticogenesis (Zhang et al. 1998). In addition, the full complement of BMP receptor subunits are expressed on VZ and SVZ progenitor cells and on other glial cell types (i.e., radial glia and oligodendroglial-type II astroglial, O-2A, progenitors) that are also BMP-responsive (Gross et al. 1996; Mabie et al. 1997; Mehler et al. 1997). In this regard, we have shown for glial lineages that BMPs promote the elaboration of astrocytes from late embryonic SVZ and early postnatal cerebral cortical multipotent progenitors, O-2A progenitors and radial glia, whereas these cytokines suppress the generation of OLs and radial glia from NSCs and their more glial-restricted progeny (Fig. 1; Mehler et al. 1995, 1997; Gross et al. 1996; Mabie et al. 1997, 1999; Marmur et al. 1998b; Mehler and Gokhan 1999; Gokhan et al. 2000). During different developmental periods, the actions of the BMPs may also be modified by interactions with other signaling molecules. For example, leukemia inhibitory factor (LIF) and BMP2 exert synergistic effects in promoting the generation of astrocytes from early embryonic cerebral cortical progenitor cells through the actions of the transcriptional co-regulator, p300/CBP (Nakashima et al. 1999a). Conversely, EGF receptor activation can inhibit the actions of BMPs by MAP kinase-mediated suppression of Smad signaling (Wrana 2000). In addition, LIF can antagonize the inhibitory actions of the BMPs to promote the generation of radial glia from early embryonic cerebral cortical NSCs and the elaboration of OLs from perinatal glial-restricted progenitor species (Fig. 1; Gokhan et al. 2000). BMP receptor expression persists within the adult cerebral cortex, although the absence of significant overlap in the regional cellular expression of individual BMP receptor subunits suggests that the BMPs may be preferentially involved in regulating cellular maintenance during mature life (Zhang et al. 1998).

3 Environmental and Transcriptional Regulation of Intermediate Progenitor Species

The profiles of environmental signals, transcriptional regulatory events and progenitor cell intermediates that are involved in orchestrating the early and progressive elaboration of neuronal and glial lineage species from embryonic cerebral cortical NSCs are still poorly defined. Certain populations of ventral neurons and oligodendrocytes of the caudal neural tube are generated from the same progenitor domains at sequential developmental stages under the influence of the secreted morphogen, Shh (Litingtung and Chiang 2000). Within this regional CNS domain, Shh has a direct inductive role in the generation of motor neurons, and this developmental action requires the continuous presence of the morphogen during the phases of cell cycle exit and early terminal differentiation (Briscoe and Ericson 1999; Litingtung and Chiang 2000). Graded Shh signals are necessary to establish the requisite progenitor populations, to initially refine the boundaries of progenitor domains, to maintain the molecular integrity of these uncommitted cells and to subsequently direct the expression of specific classes of genes that confer neuronal subtype identity (Briscoe and Ericson 1999; Briscoe et al. 2000). Shh is also expressed within the embryonic ventral forebrain and is capable of inducing the expression of transcription factors that are present in neuronal and OL precursor species (i.e., Mash 1, Olig 1, 2; Gokhan et al. 2000; Lu et al. 2000; Warburton et al. 2000; Zhou et al. 2000). We have recently shown that both Mash 1 and Olig2, but not neuronal (β-tubulin), oligodendroglial (O4) or radial glial (brain lipid binding protein, BLBP) markers, are expressed within individual progenitor species derived from early embryonic cerebral cortical NSCs exposed to Shh (Gokhan et al. 2000).

These observations support previous experimental findings that multilineage gene expression can precede actual lineage commitment (Hu et al. 1997). In addition, they suggest that Shh may promote the elaboration of neuronal-oligodendroglial (N-O) progenitors (but not radial glia) from these early embryonic NSCs. Both Shh and its signaling receptors, Patched (Ptc) and Smoothen (Smo), are expressed in early embryonic cerebral cortical paramedian generative zones, suggesting a seminal role for this signaling pathway in lineage elaboration from NSCs (Zhu et al. 1999b; Litingtung and Chiang 2000). We have shown that, in the presence of bFGF, N-terminal forms of Shh (N-Shh) enhance the proliferation of early embryonic forebrain NSCs, and subsequently promote the expression of β-tubulin while simultaneously downregulating the expression of the neuroepithelial marker, nestin, and inhibiting the expression of OL progenitors (Zhu et al. 1999b). By contrast, co-application of N-Shh and BMP2 to these dissociated NSCs abrogated the BMP-mediated anti-proliferative, OL-suppressive and later astroglial-inductive effects (Zhu et al. 1999b). Shh promotes precursor cell proliferation by activation of G1 cyclins, and sustained cell cycle progression through engagement of the G1 cyclin-retinoblastoma susceptibility gene (Rb) axis (Kenney and Rowitch 2000). Interestingly, Shh exhibits context-dependent induction of both BMPs and the BMP antagonist, noggin (Hirsinger et al. 1997; Drossopoulou et al. 2000). There are also parallel dose-response profiles for the Shh-mediated elaboration of the neuronal and oligodendroglial lineages (Briscoe et al. 2000). Shh and BMPs cooperate in the induction of ventral forebrain midline cells, and BMPs are known to exert direct effects on neural progenitor species to modify responses to Shh (Kim et al. 1997; Murtagh et al. 1999; Zhu et al. 1999b). Shh also interacts with FGF and Wnt developmental signaling pathways (Hirsinger et al. 1997; Marcelle et al. 1997; Zuniga et al. 1999). FGF and Shh exhibit cooperative effects to promote the elaboration of specific regional neuronal subtypes (Ye et al. 1998). Further, BMPs can potentiate and Shh can inhibit the activities of the Wnt signaling pathway (Marcelle et al. 1997). In addition to Shh, Wnt signaling can also potentiate the expression of the BMP antagonist, noggin (Hirsinger et al. 1997). The Wnt transduction network is involved in distinct phases of neural induction, patterning of the neural tube and in graded stages of neurogenesis (Patapoutian and Reichardt 2000). The Wnt intracellular effector molecule, disheveled, can directly inhibit the Notch lateral inhibitory pathway (Patapoutian and Reichardt 2000). Further, the Wnt-associated transcriptional modulators, LEF1/Tcf, are BMP-associated Smad co-factors (Labbe et al. 2000).

These cumulative observations suggest the presence of a complex regulatory network that may utilize bFGF, Shh, BMPs, Wnt, noggin and additional downstream effectors to orchestrate the elaboration of sequential neuronal and glial lineage species from embryonic cerebral cortical NSCs. Shh binds Ptc1 and 2 with high affinity and relieves the Ptc inhibition on Smo (Litingtung and Chiang 2000). Components of the Shh-mediated cytoplasmic signaling complex include the latent transcription factors, Gli 1–3, and the negative regulator, suppressor of fused (Ding et al. 1999; Litingtung and Chiang 2000). Gli 2 and 3 are Shh-dependent transcriptional activators or repressors with overlapping functions, whereas Gli 1 is a transcriptional activator induced by Shh-mediated Gli 2 and 3 signals (Litingtung and Chiang 2000). Shh also regulates the proteolytic processing of Gli proteins, and C-terminal Gli 2 and 3 fragments are associated with transcriptional activation, whereas N-terminal Gli fragments are associated with transcriptional repression of downstream target genes (Litingtung and Chiang 2000). N-Terminal forms of Gli 3 are preferentially bound to Smads (Liu et al. 1998). The presence of BMPs induces a dissociation of the Gli 3/Smad complex, with subsequent Smad-mediated activation of BMP-responsive genes and Gli 3-mediated antagonism of Shh activator functions of Gli 1 and 2 (Liu et al. 1998; Litingtung and Chiang

2000). Therefore, it is likely that BMP-mediated neuronal lineage elaboration from Shh-primed N-O progenitor species may involve both positive commitment to the neuronal pathway (via Smad activation) and active inhibition of OL lineage commitment (via N-terminal Gli 3 repression of transcriptional activation of target genes, such as Olig 1 and 2). Relief of the BMP-mediated inhibition and subsequent elaboration of the OL lineage may be orchestrated by expression of the BMP antagonist, noggin, through protracted exposure to Shh signaling, with sequential induction of additional positive OL maturational signals, such as PDGF and LIF (Pfeiffer et al. 1993; Miller 1996; Hirsinger et al. 1997). The induction of these later environmental cues may be necessary to overcome the inhibitory effects on OL differentiation of Notch, GGF2 and possibly ID2 and 4 (Canoll et al. 1996; Wang et al. 1998; Bongarzone et al. 2000; Gokhan et al. 2000; Wang and Barres 2000). These cooperative, complementary, sequential and antagonistic extrinsic and intrinsic regulatory influences may conspire to sculpt the evolving profiles of lineage species elaborated from early embryonic cerebral cortical NSCs.

4
Mechanisms Regulating Neuronal and Astroglial Lineage Elaboration

There are multiple pathways of neuronal differentiation, including the BMP-mediated instructive elaboration of neuronal species from early embryonic cerebral cortical NSCs (Ghosh and Greenberg 1995; Mehler et al. 1995, 1997; Johe et al. 1996; Mehler and Kessler 1996; Mayer-Proschel et al. 1997; Weiwei et al. 1998). Type III β-tubulin and a low molecular weight neurofilament isoform (NF66) are expressed following neuronal lineage commitment and prior to cell cycle exit; β-tubulin continues to be expressed in post-mitotic neurons, whereas a high molecular weight phosphorylated NF isoform (NF200) is expressed during later stages of terminal differentiation (Chiu et al. 1995; Memberg and Hall 1995; Chan et al. 1997). By contrast, both BMP-mediated transduction pathways and cytokines that signal through the LIFβ/gp130 receptor complex promote the generation of cerebral cortical astrocytes (Richards et al. 1996; Koblar et al. 1998; Rajan and McKay 1998). Targeted deletion of gp130 or the LIFβ receptor in mice results in profound reductions in glial fibrillary acidic protein (GFAP)-immunoreactive astrocytes, both in vivo and in vitro (Ware et al. 1995; Nakashima et al. 1999b). However, application of BMPs to cultures derived from gp130-/- or LIFβ receptor -/- animals results in the elaboration of astrocytes, demonstrating that the generation of astrocytes within the mammalian cerebral cortex does not require LIFβ/gp130 receptor signaling. Ciliary neurotrophic factor (CNTF) and LIF promote the generation of astrocytes from uncommitted rat embryonic forebrain progenitor cells through an instructive mechanism (Johe et al. 1996). By contrast, earlier embryonic precursor species express functional LIFβ receptors, but application of CNTF or LIF to dissociated cultures or explants failed to promote the elaboration of astrocytes (Molne et al. 2000). These observations suggest that progenitor cell responsiveness to this class of cytokines changes during early ontogeny. We have shown that at this early developmental stage, in fact, LIF promotes the elaboration of radial glia from murine NSCs by an instructive mechanism (Gokhan et al. 2000). By contrast, our data demonstrates that the subsequent generation of embryonic cerebral cortical astrocytes from radial glia requires the actions of BMPs (Gokhan et al. 2000). Our observations further reveal that this early embryonic glial developmental transition is initially inhibited both by the presence of noggin within radial glia and by the requirement for LIF-mediated induction of BMP receptors on these same radial glial species (Gokhan et al. 2000). During progressive embryonic cerebral cortical development, BMPs both enhance the generation of astrocytes from radial glia and also inhibit the elaboration of radial glia from VZ and SVZ progenitor cells, thus providing a mechanism for negative feedback regulation of radial glia by astrocytes during earlier stages of corticogenesis, and for cessation of the generation of radial glia coincident with the downregulation of neurogenesis during later stages of cerebral cortical development (Gokhan et al. 2000).

5 Developmental Regulation and Lineage Potential of Radial Glia

Bipolar radial glia, identified by expression of several cellular markers including RC2 and BLBP, initially form as a cellular subpopulation within the early murine embryonic cerebral cortical (embryonic day, E10) VZ (Edwards et al. 1990; Takahashi et al. 1990; Feng et al. 1994; Feng and Heintz 1995). We have shown that radial glia selectively respond to mitogenic signals from LIF; these proliferating species can be identified using an antibody to a phosphorylated epitope of vimentin (Kamei et al. 1998). Radial glia are present at maximal levels within forebrain regions between E14 and E16 and either disappear or become restricted as specialized regional populations during the first postnatal week (Edwards et al. 1990; Takahashi et al. 1990). Radial glia are also generated in vitro from early embryonic murine progenitor species, and can undergo a progressive transition to dual RC2/GFAP and later exclusively GFAP-immunoreactive species (Culican et al. 1990). Although developmental studies suggest that radial glia are precursors to astrocytes, several additional embryonic and perinatal glial progenitor populations are known to exist that can also give rise to astrocytes (Gray and Sanes 1992; Rao 1999; Thomas et al. 2000).

Recent studies have suggested that radial glia may, in fact, exhibit broader neural lineage potential, and may be capable of generating neuronal precursors during the period of early embryonic neurogenesis (Gray and Sanes 1992; Chanas et al. 2000; Malatesta et al. 2000). Additional investigations have further suggested that lineage species with the characteristics of astrocytes may function as NSCs within the adult brain (Doetsch et al. 1999). These developmental issues are particularly important because of the recent experimental findings that mature astrocytes can transform into transitional radial glia to support the migration of transplanted immature neurons into areas of CNS injury (Leavitt et al. 1999). However, further refinement in astroglial lineage studies has been hampered by the paucity of markers of committed lineage species. One potential early astroglial lineage marker, the Yb subunit of glutathione-S-transferase (Yb-GST), is expressed exclusively on white and gray matter astrocytes and also on evolving radial glia within subependymal regions during early CNS developmental stages (Cammer and Zhang 1992).

6
Biology of Glial-Restricted Progenitors and the Generation of Oligodendrocytes

The elaboration of the oligodendroglial lineage can also be modulated by BMPs during embryonic cerebral cortical development. OLs can be generated from embryonic forebrain NSCs by increasing concentrations of bFGF or by brief exposure to thyroid hormone (Johe et al. 1996; Qian et al. 1997). Perinatal EGF-responsive SVZ multipotent progenitors can also give rise to myelinating OLs in vivo (Hammang et al. 1997). Further, OLs can arise from glial-restricted progenitor species (Pfeiffer et al. 1993; Miller 1996). OL lineage commitment is heralded by cellular immunoreactivity to the O4 antibody, whereas the Ranscht monoclonal antibody recognizes later OL species at the time of cell cycle exit (Bansal et al. 1989; Pfeiffer et al. 1993; Miller 1996). Post-mitotic OLs are immunoreactive for the galactocerebroside epitope recognized by the GC/O1 antibody, and myelinating OLs express myelin basic protein (MBP; Pfeiffer et al. 1993; Miller 1996). In the early postnatal cerebral cortex, precursor species that spontaneously differentiate into OLs in vitro and into myelinating OLs when transplanted into the normal brain are termed O-2A progenitors (Espinosa et al. 1993; Franklin and Blakemore 1995; Miller 1996); these intermediate progenitors retain the ability to generate GFAP + /A2B5 + (type II) astrocytes under specific culture conditions (including application of BMPs) and following transplantation into CNS regions of adult hosts denuded of glial elements (Franklin and Blakemore 1995; Mabie et al. 1997). OLs can also arise from undifferentiated cells present within the SVZ and cerebral cortex proper that are distinct from O-2A progenitors, and whose differentiation is driven by distinct environmental cues present within the post-migratory microenvironment (Hardy and Friedrich 1996; Levison and Goldman 1997; Thomas et al. 2000).

Platelet-derived growth factor-α receptor (PDGFαR)-immunoreactive progenitor cells are initially expressed within the SVZ of the late embryonic forebrain and acquire A2B5 and NG2 proteoglycan-immunoreactivity at the time of cellular migration (Nishiyama et al. 1996). Within perinatal cerebral cortical regions, PDGFαR and NG2 are co-expressed by glial progenitors throughout gray and white matter tracts (Nishiyama et al. 1996). In addition to their ability to give rise to differentiated OLs, NG2 + progenitor species within specific areas of the CNS retain the potential to generate astrocytes (Levine and Card 1987; Levine and Stallcup 1987). We have identified and isolated NG2-immunoreactive progenitor species within both the early embryonic cerebral cortical VZ and the late embryonic SVZ that exhibit phenotypic similarities to O-2A progenitor cells (Gokhan et al. 2000). These glial-restricted progenitors spontaneously give rise to OLs, and application of BMPs both promotes the generation of astrocytes and also actively suppresses the elaboration of OLs (Gokhan et al. 2000). Recent studies have demonstrated that the OL lineage can exhibit a remarkable degree of developmental plasticity. Both bFGF and GGF2 promote the de novo proliferation and dedifferentiation of myelinating OLs (Grinspan et al. 1996; Canoll et al. 1999). In addition, perinatal OL progenitors are capable of conversion to NSCs by the concerted actions of bFGF and BMPs, and these NSCs can subsequently differentiate into distinct neuronal and glial lineage species under the influence of specific cytokine combinations (Kondo and Raff 2000a).

7
Role of ID Genes and Proteins in BMP-Mediated Cerebral Cortical Neural Fate Decisions

Using microarray analysis, we have shown that BMPs preferentially upregulate the expression of transcripts for ID2 and induce the expression of transcripts for ID4, but not ID1 and ID3, in early (VZ) and late (SVZ) embryonic cerebral cortical NSCs in a dose- and time-dependent manner (Gokhan et al. 2000). By contrast, co-application of BMP and LIF to embryonic cortical NSCs significantly retards the BMP-mediated upregulation and induction of ID2 and ID4 transcripts (Gokhan et al. 2000). Similar profiles of ID protein expression are associated with the cytokine manipulations, as confirmed by Western blot analysis. These overall observations demonstrate that BMPs upregulate the expression of ID2 and ID4 transcripts and proteins within early and late embryonic NSCs in vitro. Because BMPs promote progenitor cell cycle exit and cellular differentiation of these progenitor species, our findings suggest that ID2 and ID4 may mediate a set of cellular processes during neural development distinct from those recently documented for ID1 and ID3. Experimental findings from ID1-/-ID3-/- mice have demonstrated that ID1 and ID3 are normally required to prevent premature neurogenesis by preserving the integrity of the NSC state (Lyden et al. 1999). By contrast, our observations, in concert with studies that demonstrate significant differences in the profiles of expression of individual IDs, suggest that ID2 and ID4 may mediate a range of later cellular events during both cerebral cortical neurogenesis and gliogenesis.

To further examine whether the BMP-mediated upregulation of ID2 and ID4 transcripts is causally linked to the profiles of BMP-mediated neural fate decisions, antisense oligonucleotides were utilized to inhibit ID expression in early embryonic NSCs. Combined in situ/immunocytochemical analysis revealed that both uncommitted NSCs (nestin-immunoreactive without expression of markers of cells committed to specific neuronal and glial lineages) and astrocytes expressed both ID2 and ID4 transcripts, whereas intermediate-stage neurons failed to express either ID2 or ID4 (Gokhan et al. 2000). By contrast, radial glia only expressed transcripts for ID4, whereas OL progenitors expressed transcripts for ID2 and ID4 (Gokhan et al. 2000). Further, both NSCs and more immature neuronal species exhibited cytoplasmic localization of both ID2 and ID4, with nuclear translocation resulting from BMP application (Gokhan et al. 2000). By contrast, radial glia only displayed cytoplasmic localization of ID4 protein with nuclear translocation following BMP exposure (Gokhan et al. 2000). Conversely, OL progenitor cells exhibited constitutive expression of ID2 and ID4 with enhanced ID nuclear localization following BMP application (Gokhan et al. 2000). Application of antisense to ID2 significantly abrogated the BMP-mediated suppression of OL species from NSCs without affecting the BMP-mediated suppression of radial glia (Gokhan et al. 2000). Conversely, exposure of NSCs to antisense to ID4 significantly prevented the BMP-mediated suppression of radial glia from NSCs without significantly altering the BMP-mediated suppression of OL species (Gokhan et al. 2000). In addition, application of antisense to ID2 but not antisense to ID4 prevented the BMP-mediated enhancement of the generation of neurons, and this abrogation of the BMP-mediated promotion of a specific pathway of cellular differentiation was further potentiated by co-application of antisense oligonucleotides to ID2 and to ID4 (Gokhan et al. 2000).

These experimental findings suggest that ID2 and ID4 mediate a subset of both the positive and negative lineage-selective actions of the BMPs on embryonic NSCs. A significant proportion of ID2-/- mice die by postnatal day 14, whereas ID4-/- mice have a normal lifespan with no gross evidence of a neurogenic phenotype and ID2-/-ID4-/- mice exhibit perinatal lethality and a poor suckling response (M. A. Israel, pers. comm.). Behaviorally, ID2-/- animals exhibited a definite neurogenic phenotype: generalized seizures, significant hindlimb clasping and also a poor suckling response (Gokhan et al. 2000). Newborn ID2-/- animals displayed moderate reductions in brain size without preferential regional alterations (Gokhan et al. 2000). Comparative examination of analogous mid-rostral coronal sections from E15.5 mice revealed that ID2-/- animals displayed moderate reductions in the size of the periventricular generative zone and accompanying cell loss and disorganization of the intermediate zone (Kornblum et al. 1998). Interestingly, histological examination (hematoxylin and eosin) of a surviving ID2-/- animal at six months of age revealed the presence of several interspersed vertical collections of abnormal, darkly stained cells that spanned all cortical layers and were not similarly found in matched wild-type animals (Gokhan et al. 2000). Recent studies of

ID2 expression within the developing forebrain have suggested that ID2 may play an important role in regional cortical columnar pattern specification (Rubenstein et al. 1999). Further analysis of the expression of neural lineage markers in the cerebral cortex of E15.5 ID2-/- animals demonstrated significant reductions in the complement of cells that express the neuroepithelial marker, nestin, and the radial glial marker, BLBP, and an enhanced complement of early neuroblasts within the SVZ (Gokhan et al. 2000). By contrast, ID2-/- mice displayed reductions in staining and disorganization of cells expressing the neuronal marker, β-tubulin, within the cortical plate (Gokhan et al. 2000). These findings are compatible with a role for ID2 in regulating both the initial timing of neuronal lineage elaboration from early embryonic NSCs and the terminal maturation of post-mitotic neurons. These preliminary observations of ID2-/-, ID4-/- and ID2-/-ID4-/- mice suggest that ID2 and ID4 mediate non-redundant and synergistic lineage-selective processes during neural development that are distinct from those mediated by ID1 and ID3.

8 ID Genes and Proteins

8.1 Regulatory Roles

Four different ID proteins (ID1–4), each coded for by distinct cellular genes, are known to exist (Norton et al. 1998). One or more of these genes and proteins are expressed by all cell lineages (Norton et al. 1998). In general, ID gene transcripts are expressed at higher levels in proliferating cells, with reductions in levels of expression within terminally differentiated and quiescent cells (Norton et al. 1998; Lyden et al. 1999). Downregulation of the expression of ID genes occurs at the time of cell cycle exit, and ectopic ID expression frequently impairs cellular differentiation (Norton et al. 1998; Lyden et al. 1999). However, there is increasing evidence that individual IDs are capable of mediating a much broader range of different cellular functions and also a spectrum of similar functions through the actions of distinct developmental mechanisms (Norton et al. 1998). For example, IDs have been shown to exhibit diverse effects on the differentiated state, including both potentiation and inhibition of differentiation, apoptosis or promotion of alternate lineage fates that are mediated by developmental stage-specific mechanisms (Yamaguchi et al. 1991; Cooper and Newburger 1998; Florio et al. 1998; Martinsen and Bronner-Fraser 1998; Norton and Atherton 1998; Norton et al. 1998; Morrow et al. 1999). In addition, individual IDs participate in cell cycle regulation through interactions with distinct cell cycle components (Lasorella et al. 1996; Norton et al. 1998; Inoue et al. 1999). Further, within post-mitotic cells, ID proteins exhibit developmental stage- and subcellular-specific patterns of expression (Norton et al. 1998; Sablitzky et al. 1998; Tzeng and de Vellis 1998). These observations suggest that IDs may be important not only for regulating the timing of neuronal lineage commitment from NSCs, but also for modulating the timing and integrity of progressive stages of neuronal and glial lineage maturation. In fact, differences in the specificities among IDs for individual bHLH proteins, and for additional target molecules that are involved in other integral cellular processes and bind to distinct promoter sites, suggest that unique profiles of ID proteins within specific lineage species may be required to fine-tune cellular differentiation through precise regulation of specific developmental gene cascades (Neuman et al. 1995; Cooper and Newburger 1998; Deed et al. 1998).

Finally, the persistence of significant levels of ID expression within cells of adult tissues suggests that this class of proteins may also participate in the maintenance of the differentiated phenotype, perhaps by active suppression of alternate lineage fates and by prevention of cell cycle reentry (Andres-Barquin et al. 1998, 1999; Norton et al. 1998). In this regard, it is interesting to note that the first genes encoding putative OL lineage-associated bHLH proteins (Olig 1 and 2) have recently been identified (Lu et al. 2000; Takebayashi et al. 2000; Zhou et al. 2000). These genes may be analogous to neuronal determination and differentiation effector genes, are inducible by cytokines that promote the generation of the OL lineage, can promote the generation of OL precursor species and are expressed within telencephalic regions (Lu et al. 2000; Zhou et al. 2000). However, unlike neurogenic bHLH genes, they continue to be expressed in the adult state, suggesting a possible role in maintenance of the mature OL phenotype (Lu et al. 2000; Zhou et al. 2000). The expression pattern of Olig 2 is earlier and broader than that of Olig 1 (Lu et al. 2000; Takebayashi et al. 2000; Zhou et al. 2000). In addition, Olig 2 both precedes the elaboration of neuronal and OL precursors and also continues to be expressed in evolving neuroblast species (Takebayashi et al. 2000). These observations suggest that Olig 2 may be involved in orchestrating the sequential elaboration of neuronal and OL species from NSCs and may also continue to modulate their evolving cellular interrelationships, whereas Olig 1 appears to be restricted solely to expression and functions within the OL lineage.

8.2 Nervous System Functions

Within the nervous system, transcripts for ID1–3 are expressed in VZ and SVZ progenitor cells and also in proliferating neuroblasts, whereas ID2 and ID4 are present in presumptive post-mitotic neurons of both the central and the peripheral nervous systems (Neuman et al. 1993; Tzeng and de Vellis 1998; Lyden et al. 1999). Targeted deletion of both ID1 and ID3, but not of individual ID genes, is associated with embryonic lethality and a significant neurogenic phenotype: reductions in neuroepithelial cell proliferation and premature neurogenesis (Norton et al. 1998; Lyden et al. 1999). These observations indicate that ID1 and ID3 exhibit redundancy of function and are normally required to regulate the timing of neurogenesis by maintaining the NSC proliferative state and by preventing premature neuronal lineage commitment

(Lyden et al. 1999). By contrast, the distinct profiles of developmental and mature expression of ID2 and ID4 within the nervous system suggest possible alternate and unique roles for these regulatory proteins in later neuronal maturation, in the regulation of alternate neural lineage fates and in phenotypic maintenance of the differentiated state (Neuman et al. 1993; Tzeng and de Vellis 1998). For example, within the Schwann cell lineage, ID1 and ID3 are induced by injury or mitogenic stimuli, whereas ID2 and ID4 promote later myelin gene expression (Thatikunta et al. 1999). These findings suggest that ID2 and ID4 can act in concert to regulate a subset of bHLH proteins distinct from those targeted by ID1 and ID3 (Neuman et al. 1993). In addition, overexpression of ID2 induces premature neurogenesis in the dorsal neural tube (Martinsen and Bronner-Fraser 1998). Further, studies of neural cell lines suggest that ID2 can also have a range of different functions during neuronal differentiation (Neuman et al. 1993). In this regard, GAP43, involved in axonal process outgrowth, has been identified as a target of ID2 (Chiaramella et al. 1996). NSCL1, expressed in migratory post-mitotic neuroblasts, induces the expression of ID2, and overexpression of NSCL1 blocks the downregulation of ID2 gene expression during retinoic acid induced neuronal differentiation (Neuman et al. 1995).

The results of NSCL1 misexpression studies suggest that this bHLH protein may prevent cell cycle reentry of post-mitotic neuroblasts (Li et al. 1999). Further, ID2 binds to hypophosphorylated forms of the Rb protein that regulate cell cycle exit and neuronal differentiation (Lasorella et al. 1996). Both ID2 and Rb are expressed in nascent neuronal species present within the intermediate zone and the cortical plate, and Rb-/- mice exhibit ectopic mitoses and apoptosis of these specific neuronal populations that are rescued in Rb-/-ID2-/- mice (Chiu et al. 1995; Lasorella et al. 2000). These genetic analyses suggest that Rb acts upstream of ID2 to stop neuroblast proliferation and to promote early neuroblast survival and/or differentiation. Additional experimental findings suggest that a parallel Myc-ID2 pathway can override these Rb-mediated neuronal maturational events (Lasorella et al. 2000). These cumulative observations suggest that ID2 is involved in a complementary series of functions during progressive stages of neuronal differentiation (Toma et al. 2000). In addition, ID2 is also expressed in a subset of mature neurons during adult life, suggesting a possible additional role for ID2 in maintenance of the differentiated neuronal phenotype (Neuman et al. 1993). Further, ID2 is expressed in astroglial and OL lineage species (Andres-Barquin et al. 1999; Sdrulla et al. 1999). ID2 is expressed preferentially in OL progenitor cells, and overexpression of ID2 (or ID4) prevents OL differentiation (Sdrulla et al. 1999; Kondo and Raff 2000b). These findings suggest that ID2 and ID4 may be components of the timer (in addition to $p27^{Kip1}$, PDGF and thyroid hormone) that regulates the profiles of OL progenitor cell cycle progression prior to the phase of terminal differentiation (Kondo and Raff 2000b). Notch can also upregulate ID2 expression and prevent OL differentiation; these and additional experimental observations further suggest that glial species possess developmental signaling pathways similar to those of the neuronal lineage (Wang et al. 1999). ID2 is expressed in astrocytes (along with ID1, 3 and 4), although its developmental function and profiles of cellular regulation are not well characterized (Andres-Barquin et al. 1999).

In addition to exhibiting functional similarities in their regulation of Schwann cell development, ID2 and ID4 also show similarities in structural gene organization (Mantani et al. 1998). The functional overlaps in developmental regulation are reflected by selective nuclear localization of ID2 and ID4 within Schwann cell precursors (Stewart et al. 1997). ID4 exhibits the most specific nervous system expression of all the IDs, and the highest expression in the adult nervous system, in a broad pattern that differs from the more restricted distribution of ID2 (Riechmann et al. 1994; Jen et al. 1997). During development, ID4 and ID3 exhibit mutually exclusive patterns of expression in neural species, further suggesting that they subserve distinct regulatory functions (Riechmann and Sablitzky 1995). The specialized role of ID4 in later developmental events is reinforced by the observation that unlike ID1–3, ID4 is unable to negatively regulate myogenesis (Melnikova et al. 1999). ID4 is also unique among the IDs in its complex degree of transcriptional regulation and in its distinctive pathway of protein degradation (Riechmann et al. 1994; Bounpheng et al. 1999). Although all of the ID genes are expressed in astroglial species, ID4 is selectively downregulated following cAMP-induced astrocyte differentiation after a brief initial phase of enhanced transcript expression (Andres-Barquin et al. 1998, 1999). Induction of ID4 in cultured mature astrocytes results in apoptosis (Andres-Barquin et al. 1999). These experimental findings suggest that ID4 may be essential for regulating the interplay between astrocyte differentiation and apoptosis.

9
Summary and Future Directions

Although there has been a virtual revolution in our understanding of the molecular and cellular mechanisms that regulate neuronal and glial lineage commitment and terminal differentiation from NSCs during distinct stages of cerebral cortical development, detailed characterization of the dynamic interplay of regional environmental cues, cellular signaling pathways and transcriptional regulators in orchestrating these seminal maturational processes is still in its early phases of experimental analysis. Recent studies have further confirmed that NSCs exhibit sequential changes in their environmental responsiveness during distinct developmental transitions, and that specific classes of cytokines display considerable versatility in their ability to impart both inductive and inhibitory stage-specific lineage signals. Each important developmental period is characterized by the presence of converging sets of complementary, cooperative and antagonistic environmental cues and transcriptional regulatory signals that provide the necessary combinatorial and associated feedback regulatory pathways to ensure the precise timing and fidelity of cerebral cortical neurogenesis and gliogenesis and to protect against distinct forms of environmental perturbations and associated random morphogenetic errors.

The challenge of future molecular, cellular and physiological studies of cerebral cortical development will be to define the detailed signaling components, interrelated transduction cascades and progenitor cell and lineage intermediates that are involved in promoting and differentially modulating sequential stages of neurogenesis and gliogenesis from NSCs present within the embryonic and perinatal generative zone and associated forebrain regions. These biological observations will provide further insights into previously unimaginable degrees of developmental plasticity inherent in NSCs and in their more lineage-restricted progeny, and will also help to identify novel therapeutic targets for the application of innovative neural regenerative and neuroprotective paradigms using evolving gene and stem cell technologies.

Acknowledgements. I am grateful to J.A. Kessler, S. Gokhan, S.Y. Yung, P.C. Mabie and G. Zhu for continuing support and critical commentaries. Experimental studies cited from the author's laboratory were supported by grants from the NINDS, NIH and from the Kirby and Alpern Family Foundations.

References

Ahmed S, Reynolds BA, Weiss S (1995) BDNF enhances the differentiation but not the survival of CNS stem cell-derived neuronal precursors. J Neurosci 15:5765–5778

Andres-Barquin P, Hernandez MC, Israel MA (1998) Injury selectively down-regulates the gene encoding for the Id4 transcription factor in primary cultures of forebrain astrocytes. NeuroReport 9:4075–4080

Andres-Barquin PJ, Hernandez MC, Israel MA (1999) Id4 Expression induces apoptosis in astrocytic cultures and is down-regulated by activation of the cAMP-dependent signal transduction pathway. Exp Cell Res 247:347–355

Bae S, Bessho Y, Hojo M, Kageyama R (2000) The bHLH gene Hes6, an inhibitor of Hes1, promotes neuronal differentiation. Development 127:2933–2943

Bansal R, Warrington AE, Gard AL, Ranscht B, Pfeiffer SE (1989) Multiple and novel specificities of monoclonal antibodies O1, O4 and R mAb used in the analysis of oligodendrocyte development. J Neurosci Res 24:248–257

Bongarzone ER, Byravan S, Givogri MI, Schonmann V, Campagnoni AT (2000) Platelet-derived growth factor and basic fibroblast growth factor regulate cell proliferation and the expression of notch-1 receptor in a new oligodendrocyte cell line. J Neurosci Res 62:319–328

Bounpheng MA, Dimas JJ, Dodds SG, Christy BA (1999) Degradation of Id proteins by the ubiquitin-proteasome pathway. FASEB J 13:2257–2264

Bounpheng MA, Melnikova IN, Dimas JJ, Cristy BA (1999) Identification of a novel transcriptional activity of mammalian Id proteins. Nucleic Acids Res 27:1740–1746

Briscoe J, Ericson J (1999) The specification of neuronal identity by graded sonic hedgehog signalling. Cell Dev Biol 10:353–362

Briscoe J, Pierani A, Jessell TM, Ericson J (2000) A homeodomain code specifies progenitor cell identity and neuronal fate in the ventral neural tube. Cell 101:435–445

Burrows RC,Wancio D, Levitt P, Lillien L (1997) Response diversity and the timing of progenitor cell maturation are regulated by developmental changes in EGFR expression in the cortex. Neuron 19:251–267

Cameron RS, Rakic P (1991) Glial cell lineage in the cerebral cortex: a review and synthesis. GLIA 4:124–137
Cammer W, Zhang H (1992) Localization of mu class glutathione-S-transferase in the forebrains of neonatal and young rats: implications for astrocyte development. J Comp Neurol 321:33–39
Canoll PD, Musacchio JM, Hardy R, Reynolds R, Marchionni MA, Salzer JL (1996) GGF/neuregulin is a neuronal signal that promotes the proliferation and survival and inhibits the differentiation of oligodendrocyte progenitors. Neuron 17:229–243
Canoll PD, Kraemer R, Teng KK, Marchionni MA, Salzer JL (1999) GGF/neuregulin induces a phenotypic reversion of oligodendrocytes. Mol Cell Neurosci 13:79–94
Caviness VS Jr, Takahashi T (1995) Proliferative events in the cerebral ventricular zone. Brain Dev 17:159–163
Chan SO, Peng D, Chiu FC (1997) Heterogeneous expression of neurofilament proteins in forebrain and cerebellum during development: clinical implications for spinocerebellar ataxia. Brain Res 775:107–118
Chanas-Sacre G, Rogister B, Moonen G, Leprince P (2000) Radial glia phenotype: origin, regulation, and transdifferentiation. J Neurosci Res 61:357–363
Chiaramello A, Neuman T, Peavy DR, Zuber MX (1996) The GAP-43 gene is a direct downstream target of the basic helix-loop-helix transcription factors. J Biol Chem 271:22035–22043
Chiu FC, Feng L, Chan SO, Padin C, Federoff HJ (1995) Expression of neurofilament proteins during retinoic acid-induced differentiation of P19 embryonal carcinoma cells. Mol Brain Res 30:77–86
Ciccolini F, Svendsen CN (1998) Fibroblast growth factor 2 (FGF-2) promotes acquisition of epidermal growth factor (EGF) responsiveness in mouse striatal precursor cells: identification of neural precursors responding to both EGF and FGF-2. J Neurosci 18:7869–7880
Cooper CL, Newburger PE (1998) Differential expression of Id genes in multipotent myeloid progenitor cells: Id-1 is induced by early- and late-acting cytokines while Id-2 is selectively induced by cytokines that drive terminal granulocytic differentiation. J Cell Biochem 71:277–285
Culican SM, Baumrind NK, Yamamoto M, Pearlman AL (1990) Cortical radial glia: identification in tissue culture and evidence for their transformation to astrocytes. J Neurosci 10:684–692
Deed RW, Jasiok M, Norton JD (1998) Lymphoid-specific expression of the Id3 gene in hematopoietic cells. J Biol Chem 273:8278–8286
Ding Q, Fukami SI, Meng X, Nishizaki Y, Zhang X, Sasaki H, Dlugosz A, Nakafuku M, Hui C (1999) Mouse suppressor of fused is a negative regulator of sonic hedgehog signaling and alters the subcellular distribution of Gli1. Curr Biol 9:1119–1122
Doetsch F, Caille I, Lim DA, Garcia-Verdugo JM, Alvarez-Buylla A (1999) Subventricular zone astrocytes are neural stem cells in the adult mammalian brain. Cell 97:703–716
Drossopoulou G, Lewis KE, Sanz-Ezquerro JJ, Nikbakht N, McMahon AP, Hofmann C, Tickle C (2000) A model for anteroposterior patterning of the vertebrate limb based on sequential long- and short-range Shh signalling and Bmp signalling. Development 127:1337–1348
Ebendal T, Bengtsson H, Soderstrom S (1998) Bone morphogenetic proteins and their receptors: potential functions in the brain. J Neurosci Res 51:139–146
Edwards MA, Yamamoto M, Caviness VS (1990) Organization of radial glia and related cells in the developing murine CNS, an analysis based upon a new monoclonal antibody marker. Neuroscience 36:121–144
Espinosa A, Zhang M, DeVellis J (1993) O-2A progenitor cells transplanted into the neonatal rat brain develop into oligodendrocytes but not astrocytes. Proc Natl Acad Sci USA 90:50–54
Feng L, Hatten ME, Heintz N (1994) Brain lipid-binding protein (BLBP): a novel signaling system in the developing mammalian CNS. Neuron 12:895–908
Feng L, Heintz N (1995) Differentiating neurons activate transcription of the brain lipid-binding protein gene in radial glia through a novel regulatory element. Development 121:1719–1730
Ferrer I, Alcantara S, Ballabriga J, Olive M, Blanco M, Carulla R, Rivera R, Carmona M, Berruezo M, Pitrach S, Planas A (1996) Transforming growth factor α and epidermal growth factor-receptor immunoreactivity in normal and pathological brain. Prog Neurobiol 49:99–123

Florio M, Hernandez MC, Yang H, Shu HK, Cleveland JL, Israel MA (1998) Id2 promotes apoptosis by a novel mechanism independent of dimerization to basic helix-loop-helix factors. Mol Cell Biol 18:5435–5444
Franklin RGM, Blakemore WF (1995) Glial-cell transplantation and plasticity in the O-2A lineage – implication for CNS repair. Trends Neurosci 18:151–156
Furuta Y, Piston DW, Hogan BL (1997) Bone morphogenetic proteins (BMPs) as regulators of dorsal forebrain development. Development 124:2203–2212
Gage FH, Ray J, Fisher LJ (1995) Isolation, characterization, and use of stem cells from the CNS. Annu Rev Neurosci 18:159–192
Gaiano N, Nye JS, Fishell G (2000) Radial glial identity is promoted by notch1 signaling in the murine forebrain. Neuron 26:395–404
Ghosh A, Greenberg ME (1995) Distinct role for bFGF and NT3 in the regulation of cortical neurogenesis. Neuron 15:249–252
Gokhan S, Yung SY, Kessler JA, Mehler MF (2000) Cerebral cortical neurogenesis and gliogenesis require transcriptional activation of inhibitor of differentiation (ID) 2 and 4 in neural stem cells by bone morphogenetic proteins. Ann Neurol 48:415–416
Goldman JE, Zerlin M, Newman S, Zhang L, Gensert J (1997) Fate determination and migration of progenitors in the postnatal mammalian CNS. Dev Neurosci 19:42–48
Graham A, Koentges G, Lumsden A (1996) Neural crest apoptosis and the establishment of craniofacial pattern: an honorable death. Mol Cell Neurosci 8:76–83
Gray GE, Sanes JR (1992) Lineage of radial glia in the chicken optic tectum. Development 114: 271–283
Grinspan JB, Reeves MF, Coulaloglou MJ, Nathanson D, Pleasure D (1996) Re-entry into the cell cycle is required for bFGF-induced oligodendroglial dedifferentiation and survival. J Neurosci Res 46:456–464
Gross RE, Mehler MF, Mabie PC, Zang Z, Santschi L, Kessler JA (1996) Bone morphogenetic proteins promote astroglial lineage commitment by mammalian subventricular zone progenitor cells. Neuron 17:595–606
Hammang JP, Archer DR, Duncan ID (1997) Myelination following transplantation of EGF-responsive neural stem cells into a myelin-deficient environment. Exp Neurol 147:84–95
Hardy RJ, Friedrich VL Jr (1996) Oligodendrocyte progenitors are generated throughout the embryonic mouse brain, but differentiate in restricted foci. Development 122:2059–2069
Henrique D, Adam J, Myat A, Chitnis A, Lewis J, Ish-Horowicz D (1995) Expression of a delta homologue in prospective neurons in the chick. Nature 375:787–790
Hirsinger E, Duprez D, Jouve C, Malapert P, Cooke J, Pourquie O (1997) Noggin acts downstream of Wnt and sonic hedgehog to antagonize BMP4 in avian somite patterning. Development 124:4605–4614
Hogan BL (1996) Bone morphogenetic proteins in development. Curr Opin Genet Develop 6: 432–438
Hu M, Krause D, Greaves M, Sharkis S, Dexter M, Heyworth C, Enver T (1997) Multilineage gene expression precedes commitment in the hemopoietic system. Genes Dev 11:774–785
Iantosca MR, McPherson CE, Ho S-Y, Maxwell GD (1999) Bone morphogenetic proteins 2 and 4 attenuate apoptosis in a cerebellar primitive neuroectodermal tumor cell line. J Neurosci Res 56:248–258
Inoue T, Shoji W, Obinata M (1999) MIDA1, an Id-associating protein, has two distinct DNA binding activities that are converted by the association with Id1: a novel function of Id protein. Biochem Biophys Res Commun 266:147–151
Ishibashi M, Moriyoshi K, Sasai Y, Shiota K, Nakanishi S, Kageyama R (1994) Persistent expression of helix-loop-helix factor HES-1 prevents mammalian neural differentiation in the central nervous system. EMBO J 13:1799–1805
Ishibashi M, Siew-Lan A, Shiota K, Nakanishi S, Kagyeyama R, Guillemot F (1995) Targeted disruption of mammalian hairy and enhancer of split homolog-1 (HES-1) leads to upregulation of neural helix-loop-helix factors, premature neurogenesis, and severe neural tube defects. Genes Dev 9:3136–3148

Jen Y, Manova K, Benezra R (1997) Each member of the ID gene family exhibits a unique expression pattern in mouse gastrulation and neurogenesis. Dev Dyn 208:92–106

Johe KK, Hazel TG, Muller T, Dugich-Djordjevic M, McKay R (1996) Single factors direct the differentiation of stem cells from the fetal and adult central nervous system. Genes Dev 10:3129–3140

Kalyani A, Hobson K, Rao MS (1997) Neuroepithelial stem cells from the embryonic spinal cord: isolation, characterization, and clonal analysis. Dev Biol 186:202–223

Kamei Y, Inagaki N, Nishizawa M, Tsutsumi O, Taketani Y, Inagaki M (1998) Visualization of mitotic radial glial lineage cells in the developing rat brain by Cdc2 kinase-phosphorylated vimentin. GLIA 23:191–199

Kawabata M, Miyazono K (1999) Signal transduction of the TGFβ superfamily by Smad proteins. J Biochem 125:9–16

Kenney AM, Rowitch DH (2000) Sonic hedgehog promotes G(1) cyclin expression and sustained cell cycle progression in mammalian neuronal precursors. Mol Cell Biol 20:9055–9067

Kilpatrick TJ, Bartlett PF (1995) Cloned multipotential precursors from the mouse cerebrum require FGF-2, whereas glial restricted precursors are stimulated with either FGF-2 or EGF. J Neurosci 15:3653–3661

Kilpatrick TJ, Richards LJ, Bartlett P (1995) The regulation of neural precursor cells within the mammalian brain. Mol Cell Neurosci 6:2–15

Kim Dale J, Vesque C, Lints TJ, Kuber Sampath T, Furley A, Dodd J, Placzek M (1997) Cooperation of BMP7 and SHH in the induction of forebrain ventral midline cells by prechordal mesoderm. Cell 90:257–269

Koblar SA, Turnley AM, Classon BJ, Reid KL, Ware CB, Cheema SS, Murphy M, Bartlett PF (1998) Neural precursor differentiation into astrocytes requires signaling through leukemia inhibitory factor receptor. Proc Natl Acad Sci USA 95:3178–3181

Kondo T, Raff M (2000a) Oligodendrocyte precursor cells reprogrammed to become multipotential CNS stem cells. Science 289:1754–1757

Kondo T, Raff M (2000b) The Id4HLH protein and the timing of oligodendrocyte differentiation. EMBO J 19:1998–2007

Kornblum HI, Hussain RJ, Bronstein JM, Gall CM, Lee DC, Seroogy KB (1997) Prenatal ontogeny of the epidermal growth factor receptor and its ligand, transforming growth factor alpha, in the rat brain. J Comp Neurol 380:243–261

Kornblum HI, Hussain R, Wiesen HJ, Miettinen P, Zurcher SD, Chow K, Derynck R, Werb Z (1998) Abnormal astrocyte development and neuronal death in mice lacking the epidermal growth factor receptor. J Neurosci Res 53:697–717

Koyano-Nakagawa N, Kim J, Anderson D, Kintner C (2000) Hes6 acts in a positive feedback loop with the neurogenins to promote neuronal differentiation. Development 127:4203–4216

Labbe E, Letamendia A, Attisano L (2000) Association of Smads with lymphoid enhancer binding factor 1/T cell-specific factor mediates cooperative signaling by the transforming growth factor-beta and wnt pathways. Proc Natl Acad Sci USA 97:8358–8363

Lasorella A, Iavarone A, Israel MA (1996) Id2 specifically alters regulation of the cell cycle by tumor suppressor proteins. Mol Cell Biol 16:2570–2578

Lasorella A, Noseda M, Beyna M, Iavarone A (2000) Id2 is a retinoblastoma protein target and mediates signalling by Myc oncoproteins. Nature 407:592–598

Lazar LM, Blum M (1992) Regional distribution and developmental expression of epidermal growth factor and transforming growth factor α mRNA in mouse brain by a quantitative nuclease protection assay. J Neurosci 12:1688–1697

Leavitt BR, Hernit-Grant CS, Macklis JD (1999) Mature astrocytes transform into transitional radial glia within adult mouse neocortex that supports directed migration of transplanted immature neurons. Exp Neurol 157:43–57

Levine JM, Card JP (1987) Light and electron microscopic localization of a cell surface antigen (NG2) in the rat cerebellum: association with smooth and protoplasmic astrocytes. J Neurosci 7:2711–2720

Levine JM, Stallcup WB (1987) Plasticity of developing cerebellar cells in vitro studied with antibodies against the NG2 antigen. J Neurosci 7:2721–2731

Levison SW, Goldman JE (1997) Multipotential and lineage restricted precursors coexist in the mammalian perinatal subventricular zone. J Neurosci Res 48:83–94
Li CM, Yan RT, Wang SZ (1999) Misexpression of a bHLH gene, cNSCL1, results in abnormal brain development. Dev Dyn 215:238–247
Litingtung Y, Chiang C (2000) Control of Shh activity and signaling in the neural tube. Dev Dyn 219:143–154
Liu F, Massague J, Ruiz I, Altaba A (1998) Carboxy-terminally truncated Gli3 proteins associate with Smads. Nat Genet 20:325–326
Lu QR, Yuk D, Alberta JA, Zhu Z, Pawlitzky I, Chan J, McMahon AP, Stiles CD, Rowitch DH (2000) Sonic hedgehog-regulated oligodendrocyte lineage genes encoding bHLH proteins in the mammalian central nervous system. Neuron 25:317–329
Lyden D, Young AZ, Zagzag D, Yan W, Gerald WO, Reilly R, Bader BL, Hynes RO, Zhuang Y, Manova K, Benezra R (1999) Id1 and Id3 are required for neurogenesis, angiogenesis and vascularization of tumour xenografts. Nature 401:670–677
Mabie PC, Mehler MF, Papavasiliou A, Song Q, Kessler JA (1997) Bone morphogenetic proteins induce astroglial differentiation of oligodendroglial-astroglial progenitor cells. J Neurosci 17: 4112–4120
Mabie PC, Mehler MF, Kessler JA (1999) Multiple roles of bone morphogenetic protein signaling in the regulation of cortical cell number and phenotype. J Neurosci 19:7077–7088
Malatesta P, Hartfuss E, Gotz M (2000) Isolation of radial glial cells by fluorescent-activated cell sorting reveals a neuronal lineage. Development 127:5253–5263
Mantani A, Hernandez MC, Kuo WL, Israel MA (1998) The mouse Id2 and Id4 genes: structural organization and chromosomal localization. Gene 222:229–235
Marcelle C, Stark MR, Bronner-Fraser M (1997) Coordinate actions of BMPs, Wnts, Shh and noggin mediate patterning of the dorsal somite. Development 124:3955–3963
Marmur R, Kessler JA, Zu G, Gokhan S, Mehler MF (1998a) Differentiation of oligodendroglial progenitors from cortical multipotent cells requires extrinsic signals including activation of gp130/LIFβ receptors. J Neurosci 18:9800–9811
Marmur R, Mabie PC, Gokhan S, Song Q, Kessler JA, Mehler MF (1998b) Isolation and developmental characterization of cerebral cortical multipotent progenitors. Dev Biol 204:577–591
Martens DJ, Tropepe V, van der Kooy D (2000) Separate proliferation kinetics of fibroblast growth factor-responsive and epidermal growth factor-responsive neural stem cells within the embryonic forebrain germinal zone. J Neurosci 20:1085–1095
Martinsen BJ, Bronner-Fraser M (1998) Neural crest specification regulated by the helix-loop-helix repressor Id2. Science 281:988–991
Massague J (1998) TGF-beta signal transduction. Annu Rev Biochem 67:753–791
Massague J, Blain SW, Lo RS (2000) TGFβ signaling in growth control, cancer, and heritable disorders. Cell 103:295–309
Mayer-Proschel M, Kalyani AJ, Mujtaba T, Rao MS (1997) Isolation of lineage-restricted neuronal precursors from multipotent neuroepithelial stem cells. Neuron 19:773–785
McKay R (1997) Stem cells in the central nervous system. Science 276:66–71
Mehler MF, Gokhan S (1999) Postnatal cerebral cortical multipotent progenitors: regulatory mechanisms and potential role in the development of novel neural regenerative strategies. Brain Pathol 9:515–526
Mehler MF, Kessler JA (1996) Cytokine regulation of neuronal development. Crit Rev Neurobiol 9:419–446
Mehler MF, Kessler JA (1997) Hematolymphopoietic and inflammatory cytokines in neural development. Trends Neurosci 20:357–365
Mehler MF, Marmur R, Gross R, Mabie PC, Zang Z, Papavasiliou A, Kessler JA (1995) Cytokines regulate the cellular phenotype of developing neural lineage species. Int J Dev Neurosci 13: 213–240
Mehler MF, Mabie PC, Zhang D, Kessler JA (1997) Bone morphogenetic proteins in the nervous system. Trends Neurosci 20:309–317

Mehler MF, Mabie PC, Zhu G, Gokhan S, Kessler JA (2000) Developmental changes in progenitor cell responsiveness to bone morphogenetic proteins differentially modulate progressive CNS lineage fate. Dev Neurosci 22:74–85

Melnikova IN, Bounpheng M, Schatteman GC, Gilliam D, Christy BA (1999) Differential biological activities of mammalian Id proteins in muscle cells. Exp Cell Res 247:94–104

Memberg SP, Hall AK (1995) Dividing neuron precursors express neuron-specific tubulin. J Neurobiol 27:26–43

Miller RH (1996) Oligodendrocyte origins. Trends Neurosci 19:92–96

Molne M, Studer L, Tabar L, Ting Y, Eiden M, McKay R (2000) Early cortical precursors do not undergo LIF-mediated astrocytic differentiation. J Neurosci Res 59:301–311

Morrison SJ, Shah NM, Anderson DJ (1997) Regulatory mechanisms in stem cell biology. Cell 88: 287–298

Morrison SJ, Perez SE, Qiao Z, Verdi JM, Hicks C, Weinmaster G, Anderson DJ (2000) Transient notch activation initiates an irreversible switch from neurogenesis to gliogenesis by neural crest stem cells. Cell 101:499–510

Morrow MA, Mayer EW, Perez CA, Adlam M, Siu G (1999) Overexpression of the helix-loop-helix protein Id2 blocks T cell development at multiple stages. Mol Immunol 36:491–503

Morshead CM, Reynolds BA, Craig CG, McBurney MW, Staines WA, Morasutti D, Weiss S, van der Kooy D (1994) Neuronal stem cells in the adult mammalian forebrain: a relatively quiescent subpopulation of subependymal cells. Neuron 13:1071–1082

Murtagh LC, Chyung JH, Lassar AB (1999) Sonic hedgehog promotes somitic chondrogenesis by altering the cellular response to BMP signaling. Genes Dev 13:225–237

Nakamura Y, Sakakibara S, Miyata T, Ogawa M, Shimazaki T, Weiss S, Kageyama R, Okana H (2000) The bHLH gene Hes 1 as a repressor of the neuronal commitment of CNS stem cells. J Neurosci 20:283–293

Nakashima K, Yanagisawa M, Arakawa H, Kimura N, Hisatsune T, Kawabata M, Miyazono K, Taga T (1999a) Synergistic signaling in fetal brain by STAT3-Smad1 complex bridged by p300. Science 284:479–482

Nakashima K, Wiese S, Yanagisawa M, Arakawa H, Kimura N, Hisatsune T, Yoshida K, Kishimoto T, Sendtner M, Taga T (1999b) Developmental requirement of gp130 signaling in neuronal survival and astrocyte differentiation. J Neurosci 19:5429–5434

Neuman K, Nornes HO, Neuman T (1995) Helix-loop-helix transcription factors regulate Id2 gene promoter activity. FEBS Lett 374:279–283

Neuman T, Keen A, Zuber MX, Kristjansson GI, Gruss P, Nornes HO (1993) Neuronal expression of regulatory helix-loop-helix factor Id2 gene in mouse. Dev Biol 160:186–195

Nguyen VH, Trout J, Connors SA, Andermann P, Weinberg E, Mullins MC (2000) Dorsal and intermediate neuronal cell types of the spinal cord are established by a BMP signaling pathway. Development 127:1209–1220

Nishiyama A, Lin X-H, Giese N, Heldin C-H, Stallcup WB (1996) Co-localization of NG2 proteoglycan and PDGF α-receptor on O2A progenitor cells in the developing rat brain. J Neurosci Res 43:299–314

Norton JD, Atherton GT (1998) Coupling of cell growth control and apoptosis functions of Id proteins. Mol Cell Biol 18:2371–2381

Norton JD, Deed RW, Craggs G, Sablitzky F (1998) ID helix-loop-helix proteins in cell growth and differentiation. Trends Cell Biol 8:58–65

Ohtsuka T, Ishibashi M, Gradwohl G, Nakanishi S, Guillemot F, Kageyama R (1999) Hes1 and Hes5 as Notch effectors in mammalian neuronal differentiation. EMBO J 18:2196–2207

Patapoutian A, Reichardt LF (2000) Roles of Wnt proteins in neural development and maintenance. Curr Opin Neurobiol 10:392–399

Pfeiffer SE, Warrington AE, Bansal R (1993) Oligodendrocyte and its many cellular processes. Trends Cell Biol 3:191–198

Qian X, Goderie S, Shen Q, Stern J, Temple S (1998) Intrinsic programs of patterned cell lineage in isolated vertebrate CNS ventricular zone cells. Development 125:3143–3152

Qian X, Davis AD, Goderie SK, Temple S (1997) FGF2 concentration regulates the generation of neurons and glia from multipotent cortical stem cells. Neuron 18:81–93

Rajan P, McKay RDG (1998) Multiple routes to astrocytic differentiation in the CNS. J Neurosci 18:3620–3629

Rao MS (1999) Multipotent and restricted precursors in the central nervous system. Anat Rec 257:137–148

Reynolds BA, Tetzlaff W, Weiss S (1992) A multipotent EGF-responsive striatal embryonic progenitor cell produces neurons and astrocytes. J Neurosci 12:4565–4574

Reynolds BA, Weiss S (1996) Clonal and population analyses demonstrate that an EGF-responsive mammalian embryonic CNS precursor is a stem cell. Dev Biol 175:1–13

Richards LJ, Kilpatrick TJ, Dutton R, Tan S-S, Gearing DP, Bartlett PF, Murphy M (1996) Leukemia inhibitory factor or related factors promote the differentiation of neuronal and astrocytic precursors within the developing murine spinal cord. Eur J Neurosci 2:291–299

Riechmann V, Sablitzky F (1995) Mutually exclusive expression of two dominant-negative helix-loop-helix (dnHLH) genes, ID4 and ID3, in the developing brain of the mouse suggests distinct regulatory roles of these dnHLH proteins during cellular proliferation and differentiation of the nervous system. Cell Growth Differ 6:837–843

Riechmann V, van Cruchten I, Sablitzky F (1994) The expression pattern of Id4, a novel dominant negative helix-loop-helix protein, is distinct from Id1, Id2 and Id3. Nucleic Acids Res 22: 749–755

Rubenstein JLR, Anderson S, Shi L, Miyashita-Lin E, Bulfone A, Hevner R (1999) Genetic control of cortical regionalization and connectivity. Cereb Cortex 9:524–532

Sablitzky F, Moore A, Bromley M, Deed RW, Newton JS, Norton JD (1998) Stage- and subcellular-specific expression of Id proteins in male germ and Sertoli cells implicates distinctive regulatory roles for Id proteins during meiosis, spermatogenesis, and Sertoli cell function. Cell Growth Differ 9:1015–1024

Sasai Y (1998) Identifying the missing links: genes that connect neural induction and primary neurogenesis in vertebrate embryos. Neuron 21:455–458

Sdrulla A, Wang S, Barres BA (1999) Overexpression of the Id2 protein inhibits oligodendrocyte differentiation in vitro. Soc Neurosci Abstr 25:2039

Shah NM, Marchionni MA, Isaacs I, Stroobant P, Anderson DJ (1994) Glial growth factor restricts mammalian neural crest stem cells to a glial fate. Cell 77:349–360

Shah NM, Groves AK, Anderson DJ (1996) Alternative neural crest cell fates are instructively promoted by TGFβ superfamily members. Cell 85:331–343

Shen Q, Qian X, Capela A, Temple S (1998) Stem cells in the embryonic cerebral cortex: their role in histogenesis and patterning. J Neurobiol 36:162–174

Shou J, Rim PC, Calof AL (1999) BMPs inhibit neurogenesis by a mechanism involving degradation of a transcription factor. Nat Neurosci 2:339–345

Soderstrom S, Bengtsson H, Ebendal T (1996) Expression of serine/threonine kinase receptors including the bone morphogenetic factor type II receptor in the developing and adult rat brain. Cell Tissue Res 286:269–279

Song Q, Mehler MF, Kessler JA (1998) Bone morphogenetic proteins induce apoptosis and growth factor dependence of cultured sympathoadrenal progenitor cells. Dev Biol 196:119–127

Stewart HJ, Zoidl G, Rossner M, Brennan A, Zoidl C, Nave KA, Mirsky R, Jessen KR (1997) Helix-loop-helix proteins in Schwann cells: a study of regulation and subcellular localization of Ids, REB, and E12/47 during embryonic and postnatal development. J Neurosci Res 50:684–701

Takahashi T, Misson JP, Caviness VS (1990) Glial process elongation and branching in the developing murine neocortex: a qualitative and quantitative immunohistochemical analysis. J Comp Neurol 302:15–28

Takebayashi H, Yoshida S, Sugimori M, Kosako H, Kominami R, Nakafuku M, Nabeshima Y (2000) Dynamic expression of basic helix-loop-helix Olig family members: implication of Olig2 in neuron and oligodendrocyte differentiation and identification of a new member, Olig3. Mech Dev 99:143–148

Thatikunta P, Qin W, Christy BA, Tennekoon GI, Rutkowski JL (1999) Reciprocal Id expression and myelin gene regulation in Schwann cells. Mol Cell Neurosci 14:519–528
Thomas J-L, Spassky N, Perez-Villegas EM, Olivier C, Cobos I, Goujet-Zalc C, Martinez S, Zalc B (2000) Spatiotemporal development of oligodendrocytes in the embryonic brain. J Neurosci Res 59:471–476
Threadgil DW, Flugosz AA, Hansen AA, Tennenbaum T, Lichti U, Yee D, LaMantia C, Mourton T, Herrup K, Harris RC (1995) Targeted disruption of mouse EGF receptor: effect of genetic background on mutant phenotype. Science 269:230–234
Toma JG, El-Bizri H, Barnabe-Heider F, Aloyz R, Miller FD (2000) Evidence that helix-loop-helix proteins collaborate with retinoblastoma tumor suppressor protein to regulate cortical neurogenesis. J Neurosci 20:7648–7656
Torii M, Matsuzaki F, Osumi N, Kaibuchi K, Nakamura S, Casarosa S, Guillemot F, Nakafuku M (1999) Transcription factors Mash-1 and Prox-1 delineate early steps in differentiation of neural stem cells in the developing central nervous system. Development 126:443–456
Tzeng SF, de Vellis J (1998) Id1, Id2, and Id3 gene expression in neural cells during development. GLIA 24:372–381
Vescovi AL, Reynolds BA, Fraser DD, Weiss S (1993) bFGF regulates the proliferative fate of unipotent (neuronal) and bipotent (neuronal/astroglial) EGF-generated CNS progenitor cells. Neuron 11:951–966
Wang S, Barres BA (2000) Up a notch: instructing gliogenesis. Neuron 27:197–200
Wang S, Sdrulla AD, diSibio G, Bush G, Nofziger D, Hicks C, Weinmaster G, Barres BA (1998) Notch receptor activation inhibits oligodendrocytes differentiation. Neuron 21:63–75
Wang S, Hicks C, Weinmaster G, Barres BA (1999) Does the notch pathway control the timing and location of myelination? Soc Neurosci Abstr 25:2040
Warburton D, Schwarz M, Tefft D, Flores-Delgado G, Anderson KD, Cardoso WV (2000) The molecular basis of lung morphogenesis. Mech Dev 92:55–81
Ware CB, Horowitz MC, Renshaw BR, Hunt JS, Liggit D, Koblar SA, Gliniak BC, McKenna HJ, Papayannopoulou T, Thoma B, Cheng L, Donovan PJ, Peschon JJ, Bartlett PF, Willis CR, Wright BD, Carpenter MK, Davison BL, Gearing DP (1995)Targeted disruption of the low-affinity leukemia inhibitory factor receptor gene causes placental, skeletal, neural and metabolic defects and results in perinatal death. Development 121:1283–1299
Weiss S, Reynolds BA, Vescovi AL, Morshead C, Craig CG, van der Kooy D (1996a) Is there a neural stem cell in the mammalian forebrain? Trends Neurosci 9:387–393
Weiss S, Dunne C, Hewson J, Wohl C, Wheatley M, Peterson AC, Reynolds BA (1996b) Multipotent CNS stem cells are present in the adult mammalian spinal cord and ventricular neuroaxis. J Neurosci 16:7599–7609
Weiwei L, Cogswell CA, LoTurco JJ (1998) Neuronal differentiation of precursors in the neocortical ventricular zone is triggered by BMP. J Neurosci 18:8853–8862
Whitman M (1998) Smads and early developmental signaling by the TGFβ superfamily. Genes Dev 12:2445–2462
Wrana JL (2000) Regulation of smad activity. Cell 100:189–192
Yamaguchi A, Katagiri T, Ikeda T, Wozney JM, Rosen V, Wang EA, Kohn AJ, Suda T, Yoshida S (1991) Recombinant human bone morphogenetic protein-2 stimulates osteoblastic maturation and inhibits myogenic differentiation in vitro. J Cell Biol 113:681–687
Ye W, Shimamura K, Rubenstein JL, Hynes MA, Rosenthal A (1998) FGF and Shh signals control dopaminergic and serotonergic cell fate in the anterior neural plate. Cell 93:755–766
Zhang D, Mehler MF, Song Q, Kessler JA (1998) Development of bone morphogenetic protein receptors in the nervous system and possible roles in regulating TrkC expression. J Neurosci 18:3314–3326
Zhong W, Jiang M-M, Schonemann MD, Meneses JJ, Pedersen RA, Jan LY, Jan YN (2000) Mouse *numb* is an essential gene involved in cortical neurogenesis. Proc Natl Acad Sci USA 97: 6844–6849
Zhou Q, Wang S, Anderson DJ (2000) Identification of a novel family of oligodendrocyte lineage-specific basic helix-loop-helix transcription factors. Neuron 25:331–343

Zhu G, Mehler MF, Mabie PC, Kessler JA (1999a) Developmental changes in progenitor cells responsiveness to cytokines. J Neurosci Res 56:131–145
Zhu G, Mehler MF, Zhao J, Yung SY, Kessler JA (1999b) Sonic hedgehog and BMP2 exert opposing actions on proliferation and differentiation of embryonic neural progenitor cells. Dev Biol 215:118–129
Zuniga A, Haramis AP, McMahon AP, Zeller R (1999) Signal relay by BMP antagonism controls the SHH/FGF4 feedback loop in vertebrate limb buds. Nature 401:598–602

Gap Junctions and Their Implications for Neurogenesis and Maturation of Synaptic Circuitry in the Developing Neocortex

BERND SUTOR[1]

Summary. More and more data accumulate which provide evidence for an important role of gap junctions for the development and function of the mammalian brain. In the neocortex, gap junctions are already present at very early stages of development and they seem to be involved in neurogenesis and neuronal migration. At postnatal stages of development, gap junctions obviously participate in the differentiation of neurons and formation of synapses. Recently, it has been shown that they are responsible for the synchronization of inhibitory network activity, even within the adult mammalian neocortex. Gap junction-mediated interneuronal communication seems to be complementary to the signal transfer created by chemical synapses and, in some cases, these two systems interact or act synergistically. There are, however, many open questions which need to be answered before we can achieve a comprehensive understanding of the function of gap junctions and electrical synapses for the development and function of the neocortex.

1 Introduction

The mature mammalian neocortex consists of six horizontal laminae each characterized by specific types of neurons as well as by specific sets of afferent and efferent fibers. Layers II, III, V and VI contain predominantly projection neurons (pyramidal cells) with long axons leaving the neocortex and projecting to different brain regions, including other areas of the neocortex. These neurons are excitatory and use the amino acid L-glutamate as a neurotransmitter (Parnavelas 1990). In addition to these principal neurons, intrinsic interneurons or local circuit neurons of variable morphology are present in all cortical layers. Local circuit neurons are mainly inhibitory using γ-aminobutyric acid (GABA) as a neurotransmitter (Parnavelas 1990). Lamina IV represents the main input relay station of the neocortex and, therefore, it is most prominently developed in primary sensory areas. Layer I is characterized by a low cell density and by large bundles of tangentially oriented neurites originating from apical dendrites of layer II/III and layer V pyramidal cells as well as from axons of cortical and subcortical neurons projecting to the neocortex.

Complementary to this horizontal organization, vertically oriented modules exist, designated as columns (Mountcastle 1957; Goldman and Nauta 1977; Hubel and Wiesel 1977), which are interconnected by horizontal fibers (Gilbert

[1] Institute of Physiology, University of Munich, Pettenkoferstrasse 12, 80336 Munich, Germany

Results and Problems in Cell Differentiation, Vol. 39
C.F. Hohmann (Ed.): Cortical Development

1985). These modules span the neocortex from the white matter to the pial surface and they most probably represent the functional units of the neocortex. The neocortical columns seem to contain a more or less constant number of cells (Rockel et al. 1980) assembled into intricate neuronal networks (Szenthàgothai 1978) that provide the basis for the higher integrative and cognitive functions of the mammalian brain. Neocortical columns and networks result from complex developmental processes comprising, inter alia, genesis of glial cells and neurons, cell migration, cell differentiation, activity-dependent formation and elimination of synapses and apoptosis (Jacobson 1991).

In recent years, experimental data have accumulated which provide evidence for the existence of a non-synaptic neuronal communication system within the immature neocortex (for review, see Rörig and Sutor 1996a; Roerig and Feller 2000). In certain developmental periods, cells are aggregated to form functional syncytia. In these neuronal clusters, the cells are coupled to each other via gap junctions. These tunnel proteins establish a direct cytoplasmic connection between two closely apposed cells (Kumar and Gilula 1996). Gap junction-coupled functional syncytia are found not only at very early embryonic stages of neocortical development (e.g., during the period of neurogenesis), but also at postnatal stages, particularly during the period of synaptic circuit formation. It has been suggested that these neuronal clusters may represent "developmental blueprints for the adult functional architecture" of the neocortex (LoTurco and Kriegstein 1991; Yuste et al. 1992; Peinado et al. 1993a).

In this chapter, our current knowledge on gap junction expression in the immature neocortex will be summarized and the probable significance of neuronal gap junctions for neurogenesis and circuit formation in the neocortex will be discussed. Since the time course of ontogenesis varies between different species, the focus will be on the development of the rat neocortex.

1.1 Survey of Neocortical Development

The development of the mammalian neocortex occurs in a distinct pattern of successive processes (neurogenesis, migration, differentiation, synapse formation, etc.). These processes may overlap temporarily to various degrees. However, the overall process can be considered to be "essentially epigenetic: each [developmental] state is both the cause of future states and the result of prior states of the system." (Jacobson 1991, p. 410). Thus, for the purpose of temporal classification of gap junction expression in the immature neocortex and in order to assess the developmental impact of gap junction-mediated aggregation of neurons into cell clusters, a short overview of the sequence of developmental events known to contribute to the formation of synaptic circuits within the neocortex will be given (for extensive reviews, see Rakic 1982; Jacobson 1991; Parnavelas 2000).

The development of synaptic networks in the neocortex can be divided roughly into two principal periods. The first period comprises generation of cells and migration of these cells to their final position within the neocortex. Simultaneously, a coarse pattern of afferent and efferent fiber connections is established. These processes are determined mainly by activity-independent mechanisms (Shatz 1990). In the rat, this developmental stage involves the prenatal period and the first postnatal week (Miller 1988). During the second principal period of neocortical development, the final synaptic connections mature. These processes require patterned neuronal activity (Goodmann and Shatz 1993) and they depend on interactive mechanisms involving growth and withdrawal of axon branches as well as synapse formation and elimination (Shatz 1990). In the rat, synapse formation occurs predominantly in the course of the second and third postnatal weeks (Miller 1988). However, neuronal responses to afferent fiber activation can be observed as early as postnatal day 1 (P1) (Armstrong-James and Fox 1988). This indicates that synaptic connections are formed at very early stages of neocortical development and morphological studies in the rat have shown that genesis of neocortical synapses starts prenatally and the first synapses are found in the marginal zone and in the subplate (for review, see Miller 1988).

1.1.1 Neurogenesis, Migration and Development of Afferents

The neurons of the neocortex develop from the neuroepithelium of the ventricular zone which lines the wall of the rostrodorsal telencephalic forebrain vesicles. Postmitotic, i.e., newborn, neurons migrate towards the pial surface of the cerebral wall, where they enter a primordial superficial layer, the so-called preplate. This plexiform layer is formed mainly by monoaminergic fibers projecting from the midbrain and pons to the telencephalon. With ongoing development, an additional transient layer is created between the preplate and the ventricular zone. This intermediate zone consists mainly of tangentially oriented fibers.

The next major step in rat corticogenesis is the formation of the cortical plate which develops within the preplate (see Jacobson 1991). At the same time, the subventricular zone, where most of the cortical glial cells are generated, becomes evident. The cortical plate splits the preplate into the superficially located marginal zone, which becomes neocortical layer I, and the subplate. In the rat, neocortical neurons are generated between embryonic day 12 (E12) and E21 (gestation time: 21 days). The first neurons to be born are cells that differentiate as Cajal-Retzius cells and other types of neurons (e.g., subplate neurons). In the rat, these cells appear at E12 and arrive at the marginal zone at E13. Due to the splitting of the preplate by the developing cortical plate, the superficial marginal zone and the subplate contain the earliest generated neurons of the neocortex. The neurons of the cortical plate are born between E15 and E21. The cortical plate, which becomes layers II–VI of the mature neocortex, develops in an "inside-first-outside-last" sequence. Neurons born between E15 and E18 migrate to the cortical plate in about 2 days and form layers V and VI of the adult rat cortex. Those cells that are generated at E19–E21 migrate through the existing neuron population of the cortical plate and reach the superficial layers within 3–10 days. These neurons form layers II and III of the mature rat neocortex. After E19, the rate of mitosis declines dramatically and the ventricular zone stops generating neurons. Within the subventricular zone, production of cells (neurons and glia cells) continues until E21. These neurons migrate to the superficial layers of the cortical plate where they arrive between P3 and P5. After E21, only glial cells are generated in the subventricular zone.

Early in development, the neurons are guided from the ventricular zone to the cortical plate by radially aligned neuroepithelial cells. At later stages, when the pathways of migration increase and when newly born cells have to migrate through already established cortical layers, they are guided by radial glial cells which span the entire thickness of the neocortex (for review, see Jacobson 1991). It is important to note that early-generated neurons migrate rapidly and almost synchronously into the cortical plate (within 2 days). Neurons born later during gestation display a slower migration rate and migration is more asynchronous (3–10 days).

Most neocortical neurons generated in the ventricular zone are glutamatergic pyramidal cells which migrate along radial glia (Parnavelas 2000). However, a small fraction of cells change their direction within the intermediate zone and migrate orthogonal to the orientation of the radial glia (O'Rourke et al. 1992, 1995). Tangential migration seems to be one of the reasons for the spatial dispersion of clonally related cortical neurons. The small fraction of neurons which are derived from the ventricular zone and which migrate tangentially seems to consist of non-pyramidal cells that express GABA as their neurotransmitter (Tan et al. 1998). The majority of neocortical non-pyramidal cells, however, arise from the lateral and medial ganglionic eminence of the ventral telencephalon. These neurons become inhibitory interneurons (GABAergic) and they use tangential migratory pathways, most probably along corticofugal fibers. These neurons first appear in the marginal zone and, later, in the intermediate zone and cortical plate (Parnavelas 2000).

The most important functional input to the neocortex originates from the thalamus. In the mature neocortex, thalamocortical afferents terminate predominantly on neurons of layer IV and lower layer III as well as on neurons of layer VI. Thalamocortical synapses are of the asymmetric type (Gray type I, i.e., excitatory) and are found mainly (up to 90%) on dendritic spines (Creutzfeldt 1983). During development, thalamic neurons are generated at the same time as cortical neurons, but the generation of thalamic cells terminates earlier than that of neocortical neurons (Creutzfeldt 1983). Thalamocortical fibers enter the immature neocortex already before the end of the migration phase of their neocortical target neurons. After a "waiting period", during which they temporarily form synaptic contacts with subplate neurons, thalamocortical afferents grow into the cortical plate and establish synapses with newly arrived neurons (Shatz 1990; Goodman and Shatz 1993). In the rat, the establishment of thalamocortical connections occurs mainly during the first postnatal week. At the same time, the incidence of gap junction-mediated neuronal coupling is high within the neocortex.

Projections of monoaminergic neurons to the cerebral cortex occur very early in development (Jacobson 1991) and many studies have provided evidence that this early monoaminergic innervation is an important determinant for the proper development of the neocortex (for reviews, see Coyle 1982; Jacobson 1991; Roerig and Feller 2000). Noradrenergic fibers originating from the locus coeruleus invade the primordial plexiform layer of the immature rat neocortex on E17. Starting from E18, they innervate the cortical plate. This is much earlier than the ingrowth of thalamocortical axons, which enter the cortical plate on P1–P4. Noradrenergic afferents project initially to layers I and VI. In the adult rat neocortex, noradrenergic fibers are found mainly in layers V and VI of all cortical areas.

Dopaminergic afferents can be detected as early as E17 in the immature rat neocortex. They have been found in the subplate and, similar to thalamocortical afferents, they "wait" before they enter the cortical plate during P1–P3. The dopaminergic innervation arises from the rostral mesencephalon and the fibers project predominantly to the prefrontal and temporal cortex.

A dense serotonergic innervation of the rat neocortex develops during the first postnatal month. Serotonergic afferents originate from neurons of the mesencephalic raphe nuclei. A close anatomical alignment of serotonergic and thalamocortical fiber systems exists in the immature rodent neocortex suggesting an important role for serotonin during formation of thalamocortical circuits (for review, see Roerig and Feller 2000).

The trophic function of these monoaminergic afferent systems might be mediated, at least in part, via the modulation of gap junctions. It has been demonstrated that noradrenaline, dopamine and serotonin are able to modulate the permeability of neuronal gap junctions in the immature rat neocortex (Rörig et al. 1995a,b; Rörig and Sutor 1996b).

1.1.2 Development of Functional Synapses

Neuronal responses to afferent fiber activation can be observed as early as P1 (Armstrong-James and Fox 1988), indicating the presence of functional synapses at this stage of development. Even during embryonic stages, cortical evoked potentials associated with postsynaptic discharges of cortical neurons have been described (Creutzfeld 1983). Thus, development of synaptic circuits within the neocortex starts during the period of neurogenesis and neuronal migration. It is assumed that these early responses are mediated by chemical synapses. Electrophysiological investigations of the properties of ventricular zone cells provided clear evidence for the existence of important prerequisites for chemical synaptic transmission, i.e., the presence of neurotransmitters and their receptors (LoTurco and Kriegstein 1991; LoTurco et al. 1991, 1995). Neuroblasts of the ventricular zone express glutamate receptors of the AMPA and the kainate subtype. NMDA receptors do not seem to be present at this early developmental stage. In addition to glutamate receptors, $GABA_A$ receptors have been found in neuroblasts of the ventricular zone (LoTurco and Kriegstein 1991; LoTurco et al. 1991, 1995).

The first Gray type I synapses (i.e., excitatory) appear in the cortical plate between E19 and P0 (Miller 1988). Simultaneously, NMDA receptors start to emerge on cortical plate cells (LoTurco et al. 1991). Obviously, NMDA receptors are expressed within the time range lasting from the last mitotic division of cells until the arrival of the neurons at their final position in the neocortex. Tetrodotoxin-insensitive miniature excitatory postsynaptic currents (mEPSCs) can be detected as early as P3 and they consist of both a fast AMPA receptor-mediated component and a slow NMDA receptor-mediated component (Carmignoto and Vicini 1992; Burgard and Hablitz 1993b).

In addition to cortical plate neurons, subplate neurons respond to the activation of afferent and efferent fibers with the generation of excitatory postsynaptic potentials (EPSPs) and with antidromic action potentials (Friauf et al. 1990), indicating the presence of functional excitatory synapses and of ion channels necessary to generate spikes.

The first 7–10 postnatal days of rat neocortex development are characterized by the predominance of excitatory synaptic transmission (Kriegstein et al. 1987; Luhmann and Prince 1991; Agmon and O'Dowd 1992; Burgard and Hablitz 1993a,b; Kim et al. 1995). Activation of afferents evokes predominantly EPSPs in immature neocortical neurons. In addition, during the early postnatal period, immature neocortical neurons seem to be under the influence of tonically released glutamate (Kim et al. 1995). Inhibitory postsynaptic potentials (IPSPs) mediated by $GABA_A$ receptors gradually appear starting from P5–P8.

Evoked EPSPs of neonatal rats (younger than P5) display long durations, long latencies and their amplitudes decline rapidly upon stimulation with frequencies larger than 2 Hz (Burgard and Hablitz 1993a; Kim et al. 1995). Both NMDA and non-NMDA receptors are involved in the generation of these EPSPs (Agmon and O'Dowd 1992; Burgard and Hablitz 1993b; Kim et al. 1995). However, neonatal excitatory synaptic responses seem to be composed largely of NMDA receptor-dependent components (Kim et al. 1995). In the course of development, both duration and latency of excitatory synaptic potentials decrease and, by the end of the third postnatal week, evoked EPSPs resemble those observed in mature rat neocortical neurons (Carmignoto and Vicini 1992; Burgard and Hablitz 1993b; Kim et al. 1995).

The scenario described so far suggests uniform developmental processes in all areas of the neocortex. However, area-specific differences seem to exist. The kinetic properties of mEPSCs detected in immature neurons of the frontal cortex do not change significantly during the first two weeks of postnatal development (Burgard and Hablitz 1993a,b). Unitary EPSCs evoked in the same type of neurons by minimal stimulation display amplitudes and time courses similar to those of mEPSCs. Late NMDA receptor-mediated components are observed only following stronger synaptic activation. Since mEPSCs do not display such a late component and since their kinetic properties do not change with development, the authors concluded that the age-dependent decrease in EPSC/EPSP duration is not due to alterations in the gating properties of the NMDA receptor channel (Burgard and Hablitz 1993a,b). In developing neurons of the superior colliculus, the induction of tonic calcineurin activity leads to the downregulation of NMDA receptor currents (Shi et al. 2000). One might speculate that a similar mechanism is responsible for the shortening of EPSCs/EPSPs in developing rat frontal cortex neurons. In sensory areas of the neocortex, the situation seems to be different. In immature neurons of the visual cortex (Carmignoto and Vicini 1992) and the somatosensory cortex (Flint et al. 1997), the decay time constants of mEPSCs and unitary EPSCs are significantly longer than that of frontal cortical neurons and, most importantly, the decay time constants of both types of EPSCs decrease with development. It has been shown that this decrease is due to the developmental alteration in the gating properties of the NMDA receptor channel (Carmignoto and Vicini 1992; Flint et al. 1997) and that these alterations are most probably induced by changes in the subunit composition of the NMDA receptor (Flint et al. 1997).

As mentioned above, $GABA_A$ receptors are already expressed in neuroblasts of the ventricular zone and responses to exogenously applied GABA can be evoked in neocortical neurons of every developmental stage (Owens et al. 1996). At P0, GABAergic neurons are found in deep cortical layers and release of endogenous GABA from cortical slices is evident (Miller 1988). However, the release depends only weakly on the extracellular calcium concentration (Balcar et al. 1983). During an investigation of electrophysiological properties of immature rat neocortical neurons (P1–P7), we observed a significant reduction in membrane noise following application of the $GABA_A$ receptor antagonist bicuculline (Haburcak and Sutor, unpubl. obs.) suggesting that, in addition to glutamate, immature neocortical neurons seem to be affected by tonically released GABA.

Spontaneous $GABA_A$ receptor-mediated synaptic potentials in immature neocortical neurons have been described to occur at P4 (Luhmann and Prince 1991) indicating the presence of functional GABAergic synapses at these early developmental stages. However, at this time, the frequency of mIPSCs is very low (Haburcak and Sutor, unpubl. observ.) suggesting a low density of GABAergic synapses. This low density of GABAergic synapses is most probably the reason why IPSPs can hardly be evoked up to P7–P9 (Kriegstein et al. 1987; Luhmann and Prince 1991; Agmon and O'Dowd 1992; Kim et al. 1995). By the end of the first postnatal week, the number of GABAergic interneurons starts to increase gradually and it attains the mature pattern around P12–P15 (Miller 1988). Simultaneously, calcium-dependent GABA release matures and reaches adult levels at the end of the third postnatal week (Davies et al. 1975; Miller 1988). At the same time, the number of $GABA_A$

binding sites has increased by about eightfold as compared with numbers observed at early postnatal stages (Coyle 1982; Miller 1988). The development of these markers for GABAergic transmission coincides with the appearance of and increase in the number of Gray type II synapses in the neocortex (Miller 1988) and with the ability to evoke efficient $GABA_A$ receptor-mediated IPSPs in rat neocortical neurons.

Neuroblasts of the ventricular zone and embryonic neocortical neurons respond to the activation of $GABA_A$ receptors with a depolarization of the membrane potential (LoTurco et al. 1995; Owens et al. 1996). The depolarization results from a chloride equilibrium potential that is more positive than the resting membrane potential of the neurons. With maturation of chloride extrusion mechanisms, the intracellular chloride concentration declines, thereby shifting the chloride equilibrium potential to more negative values. The adult level is attained by the end of the second postnatal week (Owens et al. 1996). The GABA-induced depolarization of immature neocortical neurons is associated with an increase in the intracellular calcium concentration mediated by activation of voltage-gated calcium channels (Owens et al. 1996). It has been suggested that this so-called "excitatory" action of GABA may influence neocortical development including neurogenesis and synaptogenesis (LoTurco et al. 1991; Owens et al. 1996). However, one should keep in mind that the resting membrane potentials of immature neocortical neurons also display developmental alterations. Using the whole cell patch-clamp technique, we measured resting membrane potentials at P1/P2 with a mean value of –56 mV. At P18, this value was found to be –71 mV (Rörig et al. 1996). During the first postnatal week, the $GABA_A$ reversal potential has been determined to be close to –50 mV (Owens et al. 1996). Under these conditions, significant GABA-induced depolarizations of the membrane potential seem unlikely. Thus, one might conclude that the trophic functions of depolarizing GABA responses are restricted to embryonic neocortical neurons and that, although not fully developed, the inhibitory aspects of $GABA_A$ receptor activation is prevailing during the postnatal period of neocortical development (see also Sutor et al. 1994; Wells et al. 2000).

In summary, synaptic circuits of the neocortex start to develop during the time of neurogenesis and migration of pyramidal cells to their final positions. A consequence of the "inside-first-outside-last" pattern of cortical development is that deep layer pyramidal neurons are at their positions and form synaptic connections with other cells while those of layers II and III are still migrating. The majority of local circuit neurons (i.e., GABAergic interneurons) start to mature by the end of the first postnatal week. The period of circuit formation in the neocortex is characterized by an imbalance between inhibitory and excitatory synaptic transmission. Excitatory synaptic transmission is well expressed by the end of the embryonic period, whereas most of the inhibitory circuitry is formed during the second postnatal week. This functional imbalance creates a network instability which is reflected, for example, in an increased seizure susceptibility during that developmental period (Hablitz

1987; Sutor et al. 1994). By the end of the second postnatal week, neocortical neurons receive a strong inhibitory input. However, due to overexpression of NMDA receptors (Insel et al. 1990), these IPSPs are contaminated with strong NMDA receptor-mediated components (Luhmann and Prince 1990). These enhanced NMDA receptor-mediated responses seem to promote the activity-dependent modifications of synaptic connections.

2 Expression of Gap Junctions in the Neocortex

While chemical synaptic transmission between neocortical neurons is developing, the cells communicate via a system which is able to use two different signals for information transfer: electrical current and/or small molecules, e.g., second messengers. This intercellular communication system is created by gap junctions (for review, see Kumar and Gilula 1996). Gap junctions are formed by two plasma membrane spanning protein complexes designated as connexons or hemi-channels. A connexon consists of six subunits called connexins. Each of two closely apposed cells contributes one connexon to the gap junction. The connexons form aqueous pores with low electrical resistances. In addition, the diameters of these pores allow the diffusion of small molecules with molecular weights up to 1500 Da. Fifteen different connexins have been described. These subunits are classified according to the molecular weight of their proteins (e.g., connexin 32 or Cx32).

2.1 Expression During the Embryonic Development of the Neocortex

Patch-clamp recordings from neuroblasts of the ventricular zone revealed that these cells display membrane input resistances much lower than expected given the small diameter (≤10 μm) of these cells (LoTurco and Kriegstein 1991; Mienville et al. 1994). Injections of Lucifer Yellow, a gap junction-permeable fluorescent dye (MW 457 Da), into one cell lead to the occurrence of fluorescence signals in neighboring cells. This so-called dye coupling is taken as an indication of direct gap junction-mediated cell-to-cell connections. Pharmacological manipulations known to decrease or block gap junction permeability result in a marked increase in membrane input resistance of ventricular zone cells (LoTurco and Kriegstein 1991; Mienville et al. 1994). Thus, it has been concluded that the low membrane input resistances of ventricular zone cells are the functional consequences of intense coupling of these cells via gap junctions (LoTurco and Kriegstein 1991; Mienville et al. 1994).

The neuroblasts are coupled to build clusters of up to 90 cells. These clusters form vertically oriented columns within the ventricular zone (LoTurco and Kriegstein 1991). It has been shown that cell clustering is restricted to neural progenitor cells and radial glia (Bittman et al. 1997). The number of cells within a cluster decreases with ongoing development. At the beginning of neurogenesis in the rat neocortex (E12–E13) almost every neuroblast belongs to a cell cluster, as indicated by a dye-coupling incidence of almost 100%. Between E14 and E19, the number of clusters within the ventricular zone gradually declines and, at developmental stages close to birth (E19), no clusters are found in the ventricular zone (LoTurco and Kriegstein 1991). After E19, the ventricular zone stops generating neurons (Jacobson 1991). Thus, in the ventricular zone, the magnitude of cluster expression correlates temporarily with the degree of neurogenesis. The possible functional meaning of this correlation and the functional significance of columnar cell clusters in the ventricular zone will be discussed below.

From studies on immortalized hippocampal neuronal progenitor cell lines it is known that neuroblasts couple predominantly via gap junctions composed of Cx43 (Rozental et al. 2000). Postmitotically, the Cx43 expression in these cells declines and other connexins (e.g., Cx26, Cx33, Cx36, Cx40, and Cx45) start to appear. At later stages, when the cortical plate is formed, Cx26 seems to be the predominant connexin of the embryonic neocortex (Dermietzel et al. 1989; Nadarajah et al. 1997).

Currently, it is unknown whether subplate neurons are directly connected to each other via gap junctions and, in addition, no experimental evidence exists for gap junction coupling between migrating neocortical neurons. One might argue that complete uncoupling of cells is a prerequisite for migration. However, migration of coupled cell pairs or triads does not seem to be impossible.

2.2 Expression During the Early Postnatal Development of the Neocortex

The first postnatal week of rat neocortical development is characterized by a high incidence of gap junction-mediated coupling of pyramidal cells in deep and superficial layers (Connors et al. 1983; Peinado et al. 1993b; Rörig et al. 1996). At early stages (up to P5), gap junction-coupled cell clusters sometimes appear as columns spanning the neocortex from the developing white matter to the marginal zone (Peinado et al. 1993b; Rörig et al. 1996). With ongoing development, these cell clusters gradually disaggregate. In deep layers, uncoupling seems to commence during the second half of the first postnatal week. By the beginning of the second week, most pyramidal cells in layer V are not connected to another neuron. Only a few pairs of coupled cells can be observed (Kasper et al. 1994; Rörig et al. 1996). At the same time, concentric clusters of coupled cells are still found in layers II and III and, on average, these clusters consist of 30–40 neurons (Peinado et al. 1993b; Rörig et al. 1996). Gap junction-mediated coupling between neocortical layer II/III neurons disappears by the end of the second postnatal week (Connors et al. 1983; Peinado et al. 1993b; Rörig et al. 1996) and the incidence of coupling between pyramidal cells of the mature rat neocortex is very low (Rörig et al. 1996). However, immunocytochemical analyses and electron microscopy studies demonstrated that gap junctions are abundant in the neocortex of adult rats (Nadarajah et al. 1996). They have been identified mainly between glial cells and, less frequently, between neurons. They have also been identified between astrocytes and oligodendrocytes and, most interestingly, between glia cells and neurons. Another important finding of this study is that gap junction expression varies among different areas of the neocortex. Thus, it seems that the degree of gap junction expression depends on the specific function of cortical areas.

Using RT-PCR and Northern blotting, we recently investigated time-related changes in connexin mRNA abundance in the rat neocortex during postnatal development (Prime et al. 2000). We found that at least nine different connexins are expressed at various periods of development and that each connexin displays a characteristic developmental time course of expression. On the basis of these time courses, three groups of connexins with similar expression patterns can be classified: The first group (Cx30 and Cx45) shows no time-dependent changes throughout the first four postnatal weeks. Expression of connexins belonging to the second group (Cx26, Cx32, Cx43, and Cx46) increases during the first two postnatal weeks. The level attained at that time remains stable and is similar to that observed in the mature neocortex. The third group (Cx36, Cx37, and Cx40) shows increases in mRNA abundance during the first two postnatal weeks, followed by marked decreases in abundance during the third and fourth postnatal week. It is important to note that none of the individual time courses for connexin mRNA abundance match closely the time course for neocortical neuronal dye coupling. Many of the connexins of groups 2 and 3 show increases in mRNA abundance at a time when dye coupling is extensive and stable. The mismatch between the time courses of connexin mRNA expression and dye coupling is most probably due to the fact that expression studies performed using cortical homogenates (Prime et al. 2000) detect connexin mRNA present in all types of neocortical cells (neurons, glia and cells of blood vessels). Dye-coupling experiments provide an indicator for neuronal coupling only. In another study (Nadarajah et al. 1997), it was shown that the developmental profile of Cx26 expression during the first three postnatal weeks matches closely the development of dye coupling. Therefore, the authors suggested that neuronal coupling in the postnatal rat neocortex is mediated by Cx26. However, there are some other complementary or different possibilities which cannot be excluded: (1) a significant contribution of another connexin to neuronal coupling [e.g., the neuronal Cx36 (Condorelli et al. 2000) or unknown connexins], (2) rearrangement of connexons to form different heterotypic gap junctions, and (3) rearrangement of connexins to form different heteromeric combinations with altered dye permeability. Thus, Cx26 seems to be an important component of neuronal gap junctions expressed in neocortical neurons during early postnatal development, but there are certainly additional connexins that contribute to neuronal coupling during this period of ontogenesis. In addition to Cx26, expression in neurons has been demonstrated for Cx43 (Dermietzel et al. 1989), Cx32 (Nadarajah et al. 1996), and Cx36 (Condorelli et al. 2000).

3
Modulation of Gap Junction Permeability During Early Postnatal Stages of Neocortical Development

The permeability of gap junctions expressed in immature neocortical neurons can be modulated by a number of factors including changes in the intracellular milieu (e.g., pH_i and intracellular calcium concentration) and intracellular second messengers (for review, see Rörig and Sutor 1996a). An intracellular acidification leads to the uncoupling of immature neocortical neurons associated with a dramatic increase in neuronal input resistance and neuronal excitability (Rörig et al. 1996). Similar effects were observed upon activation of the cAMP-dependent protein kinase (Rörig et al. 1995a,b), following an increase in protein kinase C activity (Rörig and Sutor 1996b), after stimulation of inositol trisphosphate (IP_3) turnover (Rörig and Sutor 1996b) and following activation of guanylyl cyclase activity by nitric oxide (Rörig and Sutor 1996c).

The second messenger systems which modify gap junction permeability in neonatal rat neocortical neurons can be activated by a variety of neurotransmitters and/or neuromodulators known to be expressed in the immature neocortex. Monoaminergic projections to the cerebral cortex occur very early in development and evidence exists that these innervations might execute trophic functions during neocortical development. We have shown that noradrenaline as well as dopamine influence gap junction permeability via activation of the cAMP-dependent protein kinase (Rörig et al. 1995a,b). Thus, it seems possible that these neuromodulators exert their trophic function, inter alia, by modulation of gap junction permeability. Roerig and Feller (2000) have recently reviewed in depth the significance of serotonin for neocortical development.

In addition to monoamines, classical neurotransmitters like glutamate might be important for the regulation of gap junction permeability. In rat immature neurons, glutamate induces IP_3 production by activation of the metabotropic glutamate receptor (Kandler and Katz 1998). Inositol trisphosphate reduces gap junction permeability (Rörig and Sutor 1996b) and it seems to be important for the coordination of neuronal activity in the developing neocortex. The latter function requires the permeation of IP_3 through gap junctions (Kandler and Katz 1998). Surprisingly, the effects of $GABA_A$ and $GABA_B$ receptor activation on gap junction permeability have not been described so far.

4
Functional Implications of Gap Junctions in the Developing Neocortex

In recent years, convincing evidence for important functions of neuronal gap junctions in neocortical development has accumulated. Sometimes, however, it is still argued that these gap junctions represent nothing else but a rudiment of phylogenesis and that they are used as "surrogate synapses", which interconnect neurons using a primitive form of signal transfer and which substitute for chemical synapses before and during the period of circuit formation. To this argument one might reply that many of the different connexins are expressed in the developing and in the adult neocortex. During neocortical maturation, gap junctions occur earlier than chemical synapses. They are already expressed in the neuroblasts of the ventricular zone. In later periods, they coexist with chemical synapses. The studies of the developmental time courses of connexin mRNA expression in the neocortex allowed the distinction of at least three patterns: (1) stable expression throughout development and in the adult neocortex (e.g., Cx43), (2) increase in expression, particularly during postnatal development, followed by stable expression in the mature cortex (e.g., Cx32), and (3) transient high expression during the early postnatal period and decline to lower but stable levels in the adult cortex (e.g., Cx36). Our knowledge about the cell-type specific expression of the different connexins within the neocortex is still insufficient. At the moment, we cannot unambiguously assign different connexins to different cell types. However, there is evidence for the existence of functional gap junctions between certain types of neurons in the adult neocortex (Nadarajah et al. 1996; Galarreta and Hestrin 1999; Gibson et al. 1999; Fukuda and Kosaka 2000) and the lack of Cx32 in genetically altered animals correlates with the dysfunction of synaptic transmission in mature neocortical neurons (Sutor et al. 2000). Thus, the functional importance of gap junctions within the neocortex does not gradually diminish with ongoing development. Such a diminution, however, should be expected if neuronal gap junctions in the neocortex represent only rudiments of phylogenesis.

In the following sections, examples are provided which demonstrate the importance, or at least the potential importance of gap junctions for neocortical development and for the function of the mature neocortex.

4.1 Neurogenesis

In the ventricular zone, cells are coupled by gap junctions into large clusters (LoTurco and Kriegstein 1991). These clusters contain neural precursor cells and radial glia, but no differentiating and migrating neurons (Bittman et al. 1997). Due to this extensive coupling, neuroblasts forming the clusters display a low membrane input resistance. It has been estimated that 13 gap junctions contribute to the connection of two cells (LoTurco and Kriegstein 1991).

The function of gap junction-mediated cluster formation within the ventricular zone is not fully understood. However, since clusters appear predominantly during the period of neurogenesis, it seems obvious that these clusters are involved in this process. It has been suggested that the columnar organization of the clusters represents the basis of the cortical columns of the adult neocortex (LoTurco and Kriegstein 1991). At the moment, not enough experimental data exist in order to substantiate this suggestion. On the other hand, it has been shown that gap junction-mediated communication between ventricular zone cells regulates cell division (Bittman et al. 1997). Therefore, gap junction coupling of neuroblasts within the neuroepithelium seems to be important for fate determination (e.g., layer and column specificity) of neuronal progenitor cells. Coupling of neuroblasts within the ventricular zone depends on the cell cycle. During the S-phase, cells are coupled into clusters and they remain coupled through the G_2-phase. The M-phase is characterized by uncoupling and the cells recouple through the G_1-phase. In early neurogenesis, recoupling is complete by the next S-phase, whereas, at later stages of neurogenesis, the rate of recoupling declines and, during the G_2-phase, the number of uncoupled cells increases. If ventricular zone cells do not reenter the S-phase, then they do not rejoin clusters (Bittman et al. 1997). Thus, the function of gap junction-mediated coupling of neuroblasts into clusters seems to be the synchronization of the cell cycle of closely apposed and clonally related cells. At least to a certain extent, gap junction coupling of neuroblasts influences the birth dates of neurons and, in this way, their fate.

Possible signals which might control the coupling and uncoupling of ventricular zone cells are GABA and glutamate. LoTurco and Kriegstein (1991) and LoTurco et al. (1995) demonstrated the presence of $GABA_A$ receptors as well as of AMPA receptors on neuroblasts of the ventricular zone. Both transmitters induce large inward currents that result in depolarizations of the neuroblasts' membrane potentials and increases in the intracellular calcium concentration evoked mainly by activation of voltage-gated calcium channels. The large amplitudes of the membrane currents evoked by GABA and glutamate are due to the fact that the transmitters activate not only one cell, but also quite a lot of cells which are functionally coupled via gap junctions. Thus, the amplitude of these currents and, hence, their physiological effects seem to be a function of the cluster size. The GABA- and glutamate-induced depolarizations of ventricular zone cells are accompanied by a decrease in the number of cells that synthesize DNA (LoTurco et al. 1995). The authors concluded that, by depolarization of the neuroblasts' membrane potentials, the two transmitters decrease the number of the cells in the S-phase and, thereby, they reduce the cluster size. Those cells that do not rejoin the clusters start their migration into the cortical plate of the developing neocortex. Thus, a possible feedback regulation of DNA synthesis affects, directly or indirectly, the degree of gap junction-mediated cluster formation in the ventricular zone. Both the GABAergic and the glutamatergic transmitter systems differentiate early in development (Blanton and Kriegstein 1991) and migrating neurons send axons back to the superficial part of the ventricular zone (Kim et al. 1991). The growth cones of these axons might release neurotransmitters which influence cell division within a cluster of neuroblasts coupled to each other by gap junctions. GABA and glutamate are candidates for such feedback signals.

4.2
Development of Intrinsic Neuronal Properties

Resting membrane potential, input resistance and membrane time constants of rat neocortical layer II/III pyramidal cells develop during the first two to three postnatal weeks (Rörig et al. 1996). Within this period, the neurons' membrane potentials become more negative, the input resistances decrease and the membrane time constants get shorter. By the end of the third postnatal week, rat neocortical layer II/III pyramidal cells represent, at least with respect to their intrinsic neuronal properties, a fairly homogeneous neuron population (Rörig et al. 1996). Similar developmental changes have been described for the electrophysiological properties of rat layer V pyramidal cells (McCormick and Prince 1987; Kasper et al. 1994). However, in these cells, the maturation of intrinsic neuronal properties seems to be almost complete by the beginning of the second postnatal week (Kasper et al. 1994). Already at P9, the neuronal properties of layer V pyramidal neurons are not significantly different from those of the adult rat neocortex (Kasper et al. 1994). This temporal difference in the development of intrinsic neuronal properties of neurons in different neocortical layers is in good agreement with the "inside-first-outside-last" maturation of neocortical layers and it correlates with the incidence of gap junction coupling of neurons in different layers (Rörig et al. 1996).

The intrinsic properties of neocortical neurons vary markedly during the first two postnatal weeks (Rörig et al. 1996). Around P12–P14, this variability starts to decline and, by P21, it attains values similar to those observed in the adult rat neocortex. Simultaneously, the size of neuronal clusters within the neocortex, which also displays a large variability (Peinado et al. 1993b; Rörig et al. 1996), decreases and clusters disappear around P16. Thus, gap junction-mediated coupling of neurons into clusters might be important for the determination of intrinsic membrane properties. Observations made on neocortical neurons of Cx32-deficient mice might help to substantiate this hypothesis. Intracellular recordings from these neurons revealed significantly higher input resistances and shorter first equalizing time constants as compared with neurons of wild-type animals (Sutor et al. 2000). The alterations appeared to be independent of other disturbances detected in these knockout mice. These observations suggest that gap junctions (e.g., gap junctions composed of Cx32) might influence the development of intrinsic properties of neocortical neurons and, in this way, might determine the integrative properties of these cells.

4.3
Domains, Calcium Oscillations and Circuit Formation

The postnatal development of the columnar organization of neocortical circuits seems to be controlled by coordinated patterns of spontaneous neuronal activity (for reviews, see Peinado et al. 1993a; Roerig and Feller 2000). Using calcium-imaging techniques, columnar and circular domains of spontaneously coactive neurons have been described in the developing neocortex (Yuste et al. 1992). These domains occur before and persist through the main period of circuit formation in the rat neocortex (Peinado et al. 1993a).

In tangential slices of the immature neocortex, the diameters of the circular shaped domains are in the range 50–150 μm. In coronal slices of very young animals, the domains may span the entire cortex and in older rats they cover several neocortical layers. The generation of domains does not depend on the presence of action potentials and it is suppressed by compounds known to reduce gap junction permeability (Yuste et al. 1995). Thus, the neurons within such a domain are coupled to each other via gap junctions. Whether the domain size corresponds to the size of clusters formed by gap junction coupling of immature neocortical neurons is not known. It is assumed that the occurrence of domains is initialized by one or a few trigger neurons that activate the rest of the cells in a cluster via gap junctions (Peinado et al. 1993a). Recently, it has been shown that the spread of activity is mediated by a second messenger (inositol trisphosphate) permeating through gap junctions rather than by electrical current (Kandler and Katz 1998).

Whether domains of coactive neurons do play a role in circuit formation within the neocortex is, in my opinion, still hypothetical. The hypothesis is based, inter alia, on the observation that domains appear predominantly during the period of circuit formation. However, there is not only one clearly defined period in which synaptic circuits are formed within the neocortex. As outlined earlier, excitatory synaptic contacts are present before birth indicating synapse formation in the embryonic neocortex. In the deeper neocortical layers, excitatory synapses develop predominantly during the first postnatal week, whereas those of layer II/III mature mainly during the third week. Furthermore, development of inhibitory synapses starts during the second postnatal week, independent of the layer. Thus, in order to substantiate the hypothesis that domains of coactive immature neurons are required for neocortical circuit formation, it will be necessary to investigate more closely the spatiotemporal pattern of domain appearance throughout the postnatal period of neocortical development.

4.4 Electrical Coupling of Inhibitory Interneurons

In the adult rat neocortex, several different types of GABAergic local circuit neurons have been described. They differ in morphology, in electrophysiological behavior, and in the expression of calcium-binding proteins and neuropeptides (Kawaguchi 1995; Kawaguchi and Kubota 1997). The GABAergic inhibitory system controls network excitability in the neocortex. A decrease in the efficiency of neuronal inhibition leads to the induction and spread of epileptiform activity (see, e.g., Sutor et al. 1994). Since local circuit neurons represent only 10–15% of the total neuron population in the neocortex (Miller 1988), the generation of efficient inhibition requires the synchronization of their activity. Recent data has provided evidence that gap junctions might be involved in this synchronization (Galarreta and Hestrin 1999; Gibson et al. 1999). One group of GABAergic interneurons, the so-called fast-spiking cells, display a high incidence of coupling among each other (Galarreta and Hestrin 1999; Gibson et al. 1999). The coupling is cell-type-specific (i.e., no coupling with other types of cells) and, in addition, mixed connections (i.e., chemical and electrical synapses) have been detected between the same neurons. Thus, in the rat neocortex, gap junctions create a functional syncytium of inhibitory fast-spiking cells. In addition, a similar gap junction-coupled network of so-called low-threshold-spiking interneurons has been described (Gibson et al. 1999). The maturation of these networks and the development of their coupling specificity are unknown.

In mice lacking Cx32, asynchronous discharges of low-amplitude IPSPs have been observed in pyramidal neurons following activation of afferent inputs (Sutor et al. 2000). This suggests that Cx32 might be involved in the synchronization of inhibitory interneurons. Fast-spiking inhibitory interneurons express the calcium-binding protein parvalbumin (Kawaguchi 1995; Kawaguchi and Kubota 1997) and, during postnatal development of the rat neocortex, parvalbumin-expressing neurons appear around P7–P9 (Sutor et al., unpubl. obs.). With ongoing development, their density increases, reaching mature levels during the third postnatal week. Simultaneously with the appearance of parvalbumin-containing neurons in the neocortex, expression of Cx32 increases (Prime et al. 2000) and evoked IPSPs can be observed for the first time (Luhmann and Prince 1991). These temporal correlations suggest a role for Cx32 in the creation and synchronization of inhibitory networks within the neocortex.

5 Concluding Remarks

For a long time, gap junctions and electrical synapses formed by these molecular structures have been considered to be a form of inter-neuronal communication present only in invertebrates and lower vertebrates (see Bennett 2000). Now we know that gap junctions are important for the development and function of the mammalian brain (Spray and Dermietzel 1996). In the neocortex, neuronal gap junctions are involved in neurogenesis, differentiation of neurons, synapse formation and synchronization of inhibitory network activity. Gap junction-mediated interneuronal communication seems to be complementary to the signal transfer system created by chemical synapses and, in some cases, these two systems interact or act synergistically.

However, there are still problems left to solve and, with regard to the function of gap junctions in the development of the neocortex, there are many open questions. We have almost no unambiguous information about the cell-type-specific expression of connexins and its regulation during development. We do not know whether and in which cases certain types of gap junctions prefer signal transfer using electrical current or biochemical messengers. If such preferences exist, the question is whether they may change with maturation of neurons. In context with the putative functions of gap junction-coupled neuronal clusters for circuit formation within the neocortex, another complex of questions arises: How are synchronous intracellular calcium oscillations in a cluster of neurons translated into selective synapse formation on a single neuron and, are there additional second messengers (besides IP_3) involved? Furthermore, how are coupling and uncoupling regulated and does coupling/uncoupling correlate with synapse formation on a given cell? And, finally, are domains of the immature neocortex somehow related to the synaptic networks of the adult neocortex (i.e., are the domains really blueprints of the mature neocortical architecture)?

All these questions and more need to be answered before we can understand the function of gap junctions and electrical synapses for the development and function of the neocortex.

References

Agmon A, O'Dowd DK (1992) NMDA receptor-mediated currents are prominent in the thalamocortical synaptic response before maturation of inhibition. J Neurophysiol 68:345–349

Armstrong-James M, Fox K (1988) The physiology of developing cortical neurons. In: Peters A, Jones EG (eds) Cerebral cortex, vol 7. Development and maturation of cerebral cortex. Plenum Press, New York, pp 237–272

Balcar VJ, Dammash I, Wolff JR (1983) Is there a non-synaptic component in the K^+-stimulated release of GABA in the developing rat neocortex? Dev Brain Res 10:309–311

Bennett MVL (2000) Electrical synapses, a personal perspective (or history). Brain Res Rev 32: 16–28

Bittman K, Owens DF, Kriegstein AR, LoTurco JJ (1997) Cell coupling and uncoupling in the ventricular zone of developing neocortex. J Neurosci 17:7037–7044

Blanton MG, Kriegstein AR (1991) Morphological differentiation of distinct neuronal classes in embryonic turtle cerebral cortex. J Comp Neurol 310:203–210

Burgard EC, Hablitz JJ (1993a) Developmental changes in NMDA and non-NMDA receptor-mediated synaptic potentials in rat neocortex. J Neurophysiol 69:230–240

Burgard EC, Hablitz JJ (1993b) NMDA receptor-mediated components of miniature excitatory synaptic currents in developing rat neocortex. J Neurophysiol 70:1841–1852

Carmignoto G, Vicini S (1992) Activity-dependent decrease in NMDA receptor responses during development of the visual cortex. Science 258:1007–1011

Condorelli DF, Belluardo N, Trovato-Salinaro A, Mudò G (2000) Expression of Cx36 in mammalian neurons. Brain Res Rev 32:72–85

Connors BW, Benardo LS, Prince DA (1983) Coupling between neurons of the developing rat neocortex. J Neurosci 3:773–782

Coyle JT (1982) Development of neurotransmitters in the neocortex. Neurosci Res Prog Bull 20: 479–507

Creutzfeldt OD (1983) Cortex cerebri. Springer, Berlin Heidelberg New York

Davies LP, Johnston GAR, Stephenson AL (1975) Postnatal changes in the potassium-stimulated, calcium-dependent release of radioactive GABA and glycine from rat central nervous tissue. J Neurochem 23:239–259

Dermietzel R, Traub O, Hwang TK, Beyer E, Bennett MVL, Spray DC, Willecke K (1989) Differential expression of three gap junction proteins in developing and mature brain tissues. Proc Natl Acad Sci USA 86:10148–10152

Flint AC, Maisch US, Weishaupt JH, Kriegstein AR, Monyer H (1997) NR2A subunit expression shortens NMDA receptor synaptic currents in developing neocortex. J Neurosci 17:2469–2476
Friauf E, McConnell SK, Shatz CJ (1990) Functional synaptic circuits in the subplate during fetal and early postnatal development of cat visual cortex. J Neurosci 10:2601–2613
Fukuda T, Kosaka T (2000) The dual network of GABAergic interneurons linked by both chemical and electrical synapses: a possible infrastructure of the cerebral cortex. Neurosci Res 38:123–130
Galarreta M, Hestrin S (1999) A network of fast-spiking cells in the neocortex connected by electrical synapses. Nature 402:72–75
Gibson JR, Beierlein M, Connors BW (1999) Two networks of electrically coupled inhibitory interneurons in neocortex. Nature 402:75–79
Gilbert CD (1985) Horizontal integration in the neocortex. Trends Neurosci 8:160–165
Goldman PS, Nauta WJH (1977) Columnar distribution of cortico-cortical fibers in the frontal association, limbic, and motor cortex of the developing rhesus monkey. Brain Res 122:393–413
Goodman CS, Shatz CJ (1993) Developmental mechanisms that generate precise patterns of neuronal connectivity. Cell 72 / Neuron 10 [Suppl]:77–98
Hablitz JJ (1987) Spontaneous ictal-like discharges and sustained potential shifts in the developing rat neocortex. J Neurophysiol 58:1052–1065
Hubel DH, Wiesel TN (1977) Functional architecture of macaque monkey visual cortex. Proc R Soc Lond B 198:1–59
Insel TR, Miller LP, Gelhard RE (1990) The ontogeny of excitatory amino acid receptors in rat forebrain. I. *N*-Methyl-D-aspartate and quisqualate receptors. Neuroscience 35:31–43
Jacobson M (1991) Developmental neurobiology. Plenum Press, New York
Kandler K, Katz LC (1998) Coordination of neuronal activity in developing visual cortex by gap junction-mediated biochemical communication. J Neurosci 18:1419–1427
Kasper EM, Larkman AU, Lübke J, Blakemore C (1994) Pyramidal neurons in layer 5 of the rat visual cortex. II. Development of electrophysiological properties. J Comp Neurol 339:475–494
Kawaguchi Y (1995) Physiological subgroups of nonpyramidal cells with specific morphological characteristics in layer II/III of rat frontal cortex. J Neurosci 15:2638–2655
Kawaguchi Y, Kubota Y (1997) GABAergic cell subtypes and their synaptic connections in rat frontal cortex. Cereb Cortex 7:476–486
Kim GJ, Shatz CJ, McConnell SK (1991) Morphology of pioneer and follower growth cones in the developing neocortex. J Neurobiol 22:629–642
Kim HG, Fox K, Connors BW (1995) Properties of excitatory synaptic events in neurons of primary somatosensory cortex of neonatal rats. Cereb Cortex 5:148–157
Kriegstein AR, Suppes T, Prince DA (1987) Cellular and synaptic physiology and epileptogenesis of developing rat neocortical neurons in vitro. Dev Brain Res 34:161–171
Kumar NM, Gilula NB (1996) The gap junction communication channel. Cell 84:381–388
LoTurco JJ, Kriegstein AR (1991) Clusters of coupled neuroblasts in embryonic neocortex. Science 252:563–566
LoTurco JJ, Blanton MG, Kriegstein AR (1991) Initial expression and endogenous activation of NMDA channels in early neocortical development. J Neurosci 11:792–799
LoTurco JJ, Owens DF, Heath MJ, Davis MB, Kriegstein AR (1995) GABA and glutamate depolarize cortical progenitor cells and inhibit DNA synthesis. Neuron 15:1287–1298
Luhmann HJ, Prince DA (1990) Transient expression of polysynaptic NMDA receptor-mediated activity during neocortical development. Neurosci Lett 111:109–115
Luhmann HJ, Prince DA (1991) Postnatal maturation of the GABAergic system in rat neocortex. J Neurophysiol 65:247–263
McCormick DA, Prince DA (1987) Post-natal development of electrophysiological properties of rat cerebral cortical pyramidal neurones. J Physiol 393:743–762
Mienville JM, Lange GD, Barker JL (1994) Reciprocal expression of cell-cell coupling and voltage-dependent Na current during embryogenesis of rat telencephalon. Dev Brain Res 77:89–95

Miller MW (1988) Development of projection and local circuit neurons in neocortex. In: Peters A, Jones EG (eds) Cerebral cortex, vol 7. Development and maturation of cerebral cortex. Plenum Press, New York, pp 133–175

Mountcastle VB (1957) Modality and topographic properties of single neurons of cat's somatic sensory cortex. J Neurophysiol 20:408–434

Nadarajah B, Thomaidou D, Evans WH, Parnavelas JG (1996) Gap junctions in the adult cerebral cortex: regional differences in their distribution and cellular expression of connexins. J Comp Neurol 376:326–342

Nadarajah B, Jones AM, Evans WH, Parnavelas JG (1997) Differential expression of connexins during neocortical development and neuronal circuit formation. J Neurosci 17:3096–3111

O'Rourke NA, Dailey ME, Smith SJ, McConnell SK (1992) Diverse migratory pathways in the developing cerebral cortex. Science 258:299–302

O'Rourke NA, Sullivan DP, Kaznowski CE, Jacobs AA, McConnell SK (1995) Tangential migration of neurons in the developing cerebral cortex. Development 121:2165–2176

Owens DF, Boyce LH, Davis MBE, Kriegstein AR (1996) Excitatory GABA responses in embryonic and neonatal cortical slices demonstrated by gramicidin perforated-patch recordings and calcium imaging. J Neurosci 16:6414–6423

Parnavelas JG (1990) Neurotransmitters in the cerebral cortex. Prog Brain Res 85:13–29

Parnavelas JG (2000) The origin and migration of cortical neurones: new vistas. Trends Neurosci 23:126–131

Peinado A, Yuste R, Katz LC (1993a) Gap junctional communication and the development of local circuits in neocortex. Cereb Cortex 3:488–498

Peinado A, Yuste R, Katz LC (1993b) Extensive dye coupling between rat neocortical neurons during the period of circuit formation. Neuron 10:103–114

Prime G, Horn G, Sutor B (2000) Time-related changes in connexin mRNA abundance in the rat neocortex during postnatal development. Dev Brain Res 119:111–125

Rakic P (1982) Early developmental events: cell lineages, acquisition of neuronal positions, and areal and laminar development. Neurosci Res Prog Bull 20:439–451

Rockel AJ, Hiorns RW, Powell TPS (1980) The basic uniformity in structure of the neocortex. Brain 103:221–244

Roerig B, Feller MB (2000) Neurotransmitters and gap junctions in developing neural circuits. Brain Res Rev 32:86–114

Rörig B, Sutor B (1996a) Regulation of gap junction coupling in the developing neocortex. Mol Neurobiol 12:225–249

Rörig B, Sutor B (1996b) Serotonin regulates gap junction coupling in the developing rat somatosensory cortex. Eur J Neurosci 8:1685–1695

Rörig B, Sutor B (1996c) Nitric oxide stimulated increase in intracellular cGMP modulates gap junction coupling in rat neocortex. NeuroReport 7:569–572

Rörig B, Klausa G, Sutor B (1995a) Dye-coupling between pyramidal neurons in developing prefrontal and frontal cortex is reduced by protein kinase A activation and dopamine. J Neurosci 15:7386–7400

Rörig B, Klausa G, Sutor B (1995b) Beta-adrenoreceptor activation reduces dye-coupling between immature rat neocortical neurones. NeuroReport 6:1811–1815

Rörig B, Klausa G, Sutor B (1996) Intracellular acidification reduced gap junction coupling between immature rat neocortical pyramidal neurones. J Physiol Lond 490:31–49

Rozental R, Srinivas M, Gökhan S, Urban M, Dermietzel R, Kessler JA, Spray DC, Mehler MF (2000) Temporal expression of neuronal connexins during hippocampal ontogeny. Brain Res Rev 32:57–71

Shatz CJ (1990) Impulse activity and the patterning of connections during CNS development. Neuron 5:745–756

Shi J, Townsend M, Constantine-Paton M (2000) Activity-dependent induction of tonic calcineurin activity mediates a rapid developmental downregulation of NMDA receptor currents. Neuron 28:103–114

Spray DC, Dermietzel R (1996) Gap junctions in the nervous system. Landes, Austin

Sutor B, Hablitz JJ, Rucker F, ten Bruggencate G (1994) Spread of epileptiform activity in the immature rat neocortex studied with voltage-sensitive dyes and laser scanning microscopy. J Neurophysiol 72:1756–1768

Sutor B, Schmolke C, Teubner B, Schirmer C, Willecke K (2000) Myelination defects and neuronal hyperexcitability in the neocortex of connexin 32-deficient mice. Cereb Cortex 10:684–697

Szenthágothai J (1978) The neuron network of the cerebral cortex: a functional interpretation. Proc R Soc Lond B 201:219–248

Tan SS, Kalloniatis M, Sturm K, Tam PP, Reese BE, Faulkner-Jones B (1998) Separate progenitors for radial and tangential cell dispersion during development of the cerebral neocortex. Neuron 21:295–304

Wells JE, Porter JT, Agmon A (2000) GABAergic inhibition suppresses paroxysmal network activity in the neonatal rodent hippocampus and neocortex. J Neurosci 20:8822–8830

Yuste R, Peinado A, Katz LC (1992) Neuronal domains in developing neocortex. Science 257: 665-669

Yuste R, Nelson DA, Rubin WW, Katz LC (1995) Neuronal domains in developing neocortex: Mechanisms of coactivation. Neuron 14:7–17

Influence of Radial Glia and Cajal-Retzius Cells in Neuronal Migration

Marcin Gierdalski and Sharon L. Juliano[1]

Summary. Normal development of cerebral cortex depends on proper sequential genesis of cortical neurons and glia. Disruption of corticogenesis in ferret by short-term arresting of cell division using injections of methylazoxy methanol (MAM) leads to a specific constellation of effects, including disruption and early differentiation of radial glia into astrocytes and disorganization of reelin-containing Cajal-Retzius cells. We hypothesize that early interference of normal cortical development removes a factor instrumental in maintaining radial glia in their normal elongated shape. In support of this idea, coculture of MAM-treated slices with normal cortical plate restores radial glia and Cajal-Retzius cells to their normal positions. Recently, we found that conditioned medium obtained from normal organotypic cultures returned radial glia toward their normal morphology only in a fraction of 30–50 kDa molecular weight (MW). To assess whether restoring this factor would also improve effective migration into the cortical plate of E24 MAM-treated animals, we conducted experiments using cocultures of normal cortical plate with organotypic cultures of MAM-treated cortex, which received prior BrdU injections. In both the normal and E24 MAM-treated/normal cortical plate coculture, a greater percentage of BrdU positive cells migrated effectively into the cortical plate. We suggest that early interruption of cell division eliminates a population of cells and a factor important for maintaining proper cortical development, specifically providing cues maintaining elongation of radial glia.

1 Radial Glial Cells

The role of radial glia in neural development has evolved over many years; it has been recognized for more than a century that these elongated processes play an important role in the development of the cerebral cortex. It is currently acknowledged that radial glia are cells that stretch between the ventricular and pial surfaces, which serve as a source of support and direction for cells migrating into the cerebral cortex. They also possess specific properties that are clearly associated with glial cells (Levitt and Rakic 1980; Levitt et al. 1981). A detailed review by Bentivoglio and Mazzarello (1999) reports that, by the late 1800s, Kölliker and His had observed lengthened, radially placed cells that lined the ventricles. His, in particular, described germinal cells in the ventricular epithelium and probable bipolar neuroblasts that migrated along radial fibers towards the cerebral cortex. Also in the late nineteenth century, the advent of the Golgi technique allowed Magini to visualize these elongated

[1] Department of Anatomy and Cell Biology, and Program in Neuroscience, USUHS, 4301 Jones Bridge Rd, Bethesda, Maryland 20814, USA

Results and Problems in Cell Differentiation, Vol. 39
C.F. Hohmann (Ed.): Cortical Development

fibrous processes and he probably initiated the term "radial cells". Magini was one of the earliest to describe putative migrating neurons in association with the radial processes (Bentivoglio and Mazzarello 1999).

During the first part of the twentieth century, knowledge of these radially oriented cells did not expand greatly. Prior to the 1950s, varicosities or swellings were noticed along the length of the radial fibers by several researchers, which were suggested by some to be immature neurons. During the 1960s, a model presented by Berry and Rogers (1965) described the idea that radial glia may generate "neuroblasts". At that time they suggested that neuroblasts were generated within the cytoplasm of the elongated radial fibers.

By the 1970s, these cells had attracted the attention of Rakic who studied them extensively in relation to the cerebellum and cerebral cortex and confirmed their glial nature. Rakic also firmly established that neurons use radial glial fibers to guide them on their route to the cerebral cortex (Rakic 1971, 1972). He and coworkers furthermore determined that radial glial cells differentiate into astrocytes when they are no longer needed as a scaffold (Schmechel and Rakic 1979). These elongated cells are therefore of critical significance for establishing the normal laminar patterns in the cerebral cortex. In addition to this important function, several investigators have recently provided convincing evidence that radial glial cells are the source of neurons that migrate towards the cerebral cortex (Noctor et al. 2001a). These researchers, and others, have promoted the idea that cells generated from radial glia provide the basis for the overall radial and columnar organization in the cerebral cortex (Malatesta et al. 2000; Hartfuss et al. 2001; Miyata et al. 2001).

2 Cajal-Retzius Cells and Reelin

Another group of cells that have a long history, and are important in influencing cells migrating into the neocortex, are the Cajal-Retzius cells. As indicated by their name, they were recognized around the turn of the nineteenth century by both Cajal and Retzius. These cells are located in layer 1 of the cerebral cortex, and probably comprise a transient population (Marin-Padilla 1998). Several different morphologies associated with these cells have been described, and both Cajal and Retzius illustrated different associated shapes (e.g., see Meyer et al. 1999 for review). They are often considered to be bipolar, with extensive horizontal projections.

These distinctive cells have recently captured substantial attention because of their association with an interesting mutant mouse. The reeler mutant is a naturally occurring mutation in which the gene encoding the protein reelin is deleted. An aspect of the phenotype resulting from this mutation appears as a disrupted cortical laminar pattern. In some areas of the brain the cortical layers are reversed from their normal positions, so that the normally deep layers (5–6) are closest to the surface of the brain while layers usually the most superficial are positioned most deeply (Caviness and Sidman 1973). The discovery that the reeler mutant lacks a protein normally present in layer 1 of the cortex expanded the idea that Cajal-Retzius cells, and their constituent molecule reelin, are important in determining the ultimate position of cells migrating into the neocortex. Today, Cajal-Retzius cells are most clearly defined as those immunoreactive for reelin (Rice and Curran 2001). A series of experiments addressing the role of reelin during neocortical migration have provided a strong rationale for considering this molecule as an important player in this process.

Reelin is a secreted glycoprotein that is expressed by Cajal-Retzius cells, and a few other CNS cell types. Reelin is necessary for proper positioning of cells as they migrate into the neocortex, although the precise action of reelin is not defined. Reelin may attract migrating cells, or repel subplate neurons (Rice and Curran 2001). There is also a suggestion that reelin may act at a distance on target cells, providing a repulsive signal for migrating neurons (Lambert de Rouvroit and Goffinet 1998).

Although the focus of substantial research in recent years, the precise role of reelin in neocortical development is still not completely clear. In normal animals, the earliest generated cells in the neocortex are known collectively as the preplate, which stratify into the marginal zone (future layer 1) and the subplate. Newly generated neurons insinuate themselves between the marginal zone and subplate and form the well-known inside-out laminar pattern of the neocortex, with the deepest layer of the cortex (layer 6) arriving first, and the more superficial layers arriving later. In the reeler mouse, these two earliest generated layers fail to split, remain together as a "superplate", and newly arriving neurons pile up in improper positions (Fig. 1).

Several other molecules in the signaling pathway of reelin have been identified that appear to be important in mediating neuronal migration into the cerebral cortex and cerebellum. At least 17 different genes have been associated with the reelin signaling pathway and appear to affect neuronal positioning during development (for review, see Rice and Curran 2001). Although it is beyond the scope of this chapter to present these in detail, they operate on a number of different levels manifested in varying degrees of phenotypic similarity to the reeler mouse. Some examples include the genes *disabled-1*, *cdk5*, and *p35*. Mutation of the *disabled-1* gene produces characteristics similar to those in the reeler, resulting in improper positioning of cells. Howell and colleagues suggest that this gene may function when reelin binds to a cell and stimulates phosphorylation of Dab-1, which triggers a further intracellular cascade. In mice deficient in the *disabled-1* gene, although reelin is present, Dab-1 is not phosphorylated, and this leads to neuroanatomical phenotypes similar to the reeler (Howell et al. 1997a,b, 1999a,b). The *cdk5* and *p35* genes appear important in later stages of migration and cortical formation; structural and functional disruptions also occur in their absence. These genes may play a role in mediating the adhesive quality of neurons, perhaps in the later stages of migration, resulting in a defect more specific to the upper layers of

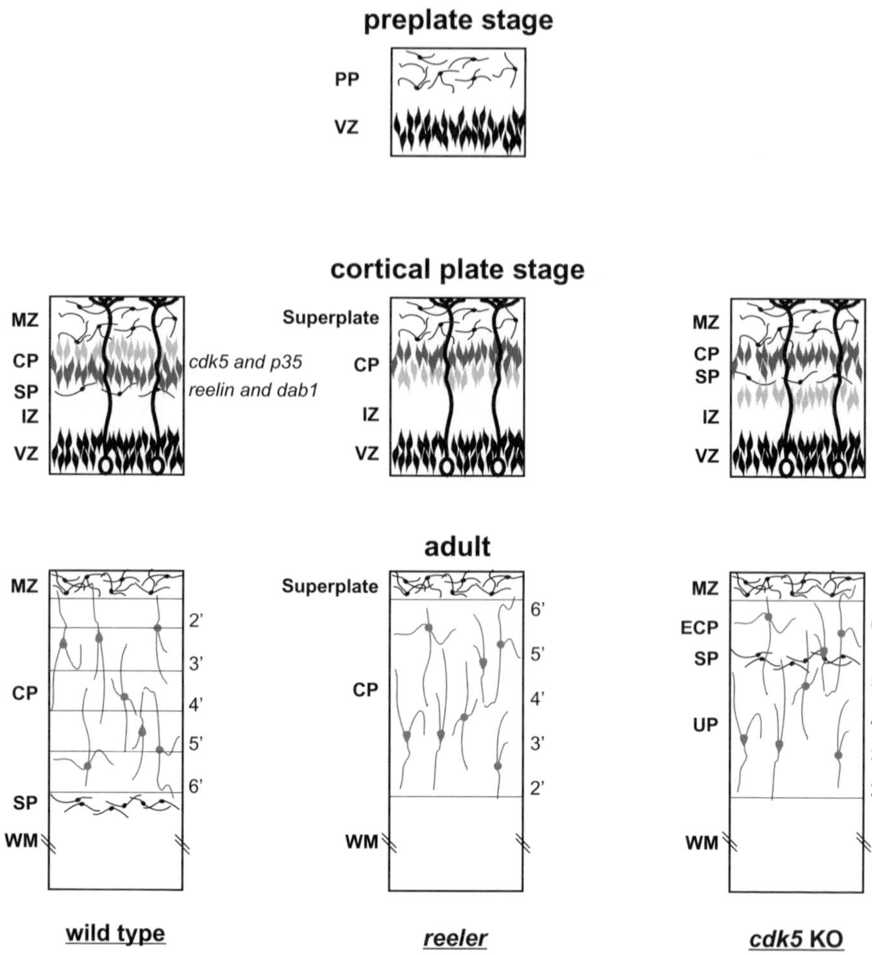

Fig. 1. Montage illustrating different stages of neocortical development in wild-type, reeler mutant and cdk5 KO (knockout) mice. The earliest stage of cortical formation (preplate stage) is shown at the *top*, which consists of the proliferative ventricular zone (*VZ*) and the preplate. The preplate normally separates into the marginal zone (*MZ*) and the subplate (*SP*) during the cortical plate stage; the newly generated neurons that form the cortical plate (*CP*) insinuate themselves between the MZ and SP. The approximate sites of action for several genes mediating cell positioning during neocortical migration are indicated on the middle left panel (*cdk5* and *P35*, *reelin* and *dab1*). The patterns of migration for each type of mouse are indicated in the *middle* and *bottom panels*. In the wild-type animals, newly generated neurons migrate on radial glial cells where they form the cortical layers (2–6) positioned between the MZ and SP. In the reeler mouse, the preplate fails to split (forming the *superplate*) and the cortical layers organize into a roughly inverted pattern. In the *cdk5* KO (knockout), the preplate splits, but only allows the first arriving layers to position themselves in the proper position above the SP, and the subsequent layers form an inverted underplate (*UP*). The subplate is located between the underplate (*UP*) and the early cortical plate (*ECP*). *WM* White matter; *IZ* intermediate zone. (Adapted from Gleeson and Walsh 2000)

neocortex (Fig. 1; Gilmore et al. 1998; Gilmore and Herrup 2000; Kwon and Tsai 2000; Kwon et al. 2000).

3 MAM Model

In our laboratory, we have been using a model that disrupts cortical development early during the period of corticogenesis. To do this, we administer an antimitotic (methylazoxy methanol; MAM) on a day when the first layers of the cerebral cortex are normally generated. MAM interrupts the birth of cells generated during a restricted period of time by methylating guanine residues in DNA. This consequently blocks the action of DNA polymerase and causes decreased DNA synthesis (Matsumoto and Higa 1966; Zedeck et al. 1970). In our hands, active mitosis resumes in 8–10 h (Noctor et al. 1999, 2001b; Palmer et al. 2001).

If this drug is delivered early during corticogenesis of the ferret, a specific constellation of effects results from the treatment. The following features can be observed on the first day of birth (P0): (1) disrupted radial glial cells, with their early differentiation into astrocytes, (2) disorganized distribution of Cajal-Retzius cells, which are normally confined to layer 1, and (3) thin and poorly laminated cerebral cortex (Noctor et al. 1999). Figure 2 shows the effect on radial glial cells after MAM injections on embryonic day 24 (E24) in pregnant ferrets. After MAM treatment, radial glial cells are no longer spoke-like; the elongated fibers are oriented in numerous directions and display early differentiation into astrocytes. This process causes the transformed radial glia to be less useful as a substrate for migration, as they no longer support a direct path to the cortical plate.

After E24 MAM treatment in ferrets, the Cajal-Retzius cells, normally precisely located in layer 1, are scattered throughout the thickness of the cortical wall (Fig. 3). In addition, the cortex is thin and poorly formed, with ectopic clusters of cells observed in the cortex and underlying intermediate zone. Taken together, these findings suggest that several features in the MAM-treated brain combine to inhibit successful migration of cells into the neocortex. This treatment disrupts two elements important for establishing a proper laminar pattern during neocortical migration: radial glia and Cajal-Retzius cells. Because of the disrupted and no longer spoke-like radial glia, newly generated neurons are unable to properly negotiate toward their intended target and wander to inappropriate sites. In addition, because the Cajal-Retzius cells are in unusual positions, outside of their normal layer 1 territory, they may be signaling to the migrating cells to stop traveling in the vicinity of the displaced cells.

One relatively normal feature in the E24 MAM-treated animals is the thalamocortical projection. Despite the severely disrupted neocortex with poor lamination, the thalamic nuclei are directed precisely to their normal targets. We noticed that the ventrobasal complex projects specifically and directly to

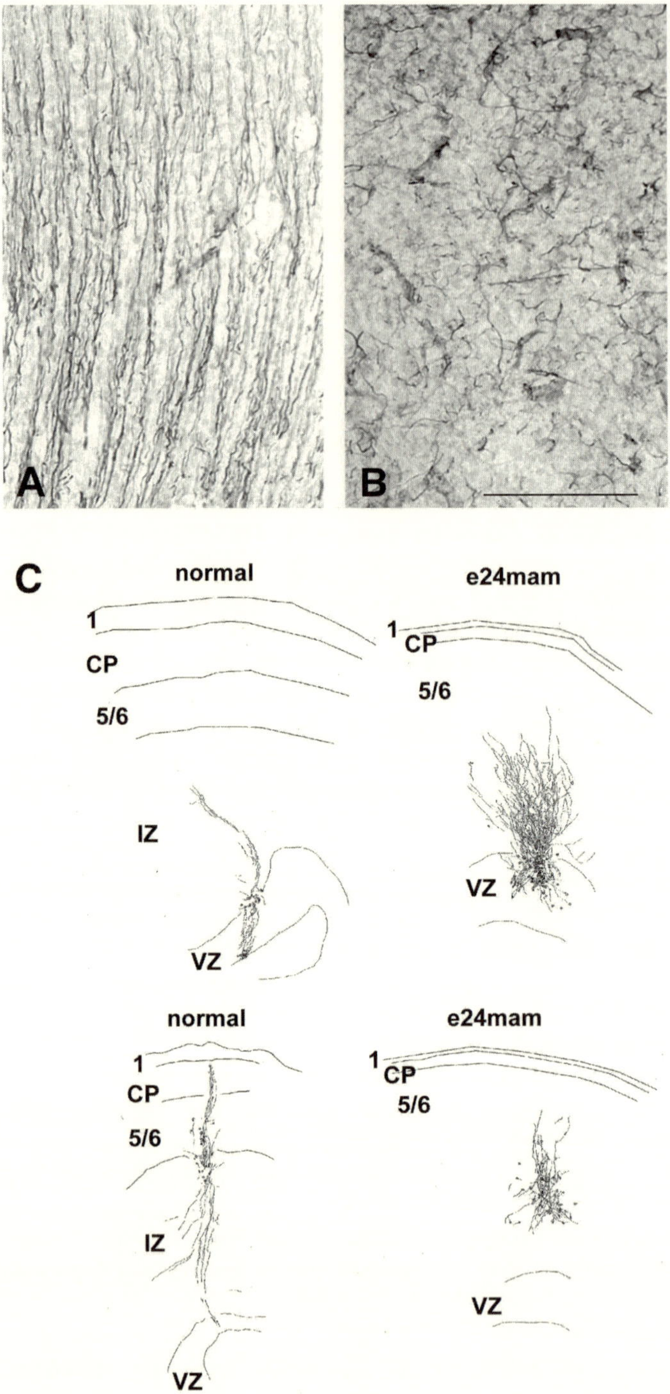
A
B
C
normal
e24mam
1
CP
5/6
IZ
VZ
1
CP
5/6
VZ
normal
e24mam
1
CP
5/6
IZ
VZ
1
CP
5/6
VZ

Fig. 2A–C

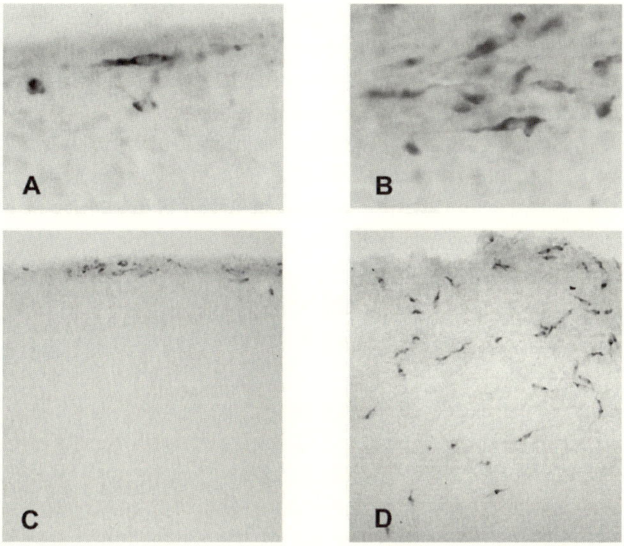

Fig. 3A–D. Examples of immunoreactivity to demonstrate the presence of reelin. **A** and **C** are from normal ferrets on P0 and demonstrate the normal position and morphology of Cajal-Retzius cells. **B** and **D** demonstrate the positions of reelin-containing Cajal-Retzius cells in E24 MAM-treated brains. In the MAM-treated brains, the reelin-positive cells are scattered throughout the cortical wall and are not restricted to layer 1. (Reprinted with permission from Noctor et al. 1999)

the parietal cortex, while the lateral geniculate nucleus projects directly to the occipital cortex. Although slightly unrelated to the focus presented here, this observation suggests that the thalamocortical projection is guided by cues other than those disrupted in our model.

In addition, the relatively recent observation that radial glia may be the source of newly generated neurons suggests that the normal pattern of neuron production may be interrupted by the altered state of the radial glia (Noctor et al. 2001a). Our earlier data indicated that neurons continue to be born after administration of MAM, but the rate of birth and the number of neurons generated may be slowed.

Fig. 2A–C. Examples of the effect of MAM injections on E24 on radial glial cells. Examples of vimentin immunoreactivity in **A** normal and **B** E24 MAM-treated sections. Vimentin is a good marker for radial glial cells in ferrets. The normal pattern of vimentin immunoreactivity is much more organized in normal sections than in MAM-treated, where the pattern is fractionated and not radially organized. **C** Drawings of labeled radial glial cells, obtained by injecting fluorescently tagged dextrans into live slices of either normal or E24 MAM-treated ferret brains. The labeled radial glia in the MAM-treated slices are more disorganized and less radial than those revealed in the normal slices. *CP* Cortical plate; *VZ* ventricular zone. (Reprinted with permission from Noctor et al. 1999)

There are similarities and distinctions between the phenotype of the reeler mouse and the E24 MAM-treated ferret. Both animal models have evidence of disrupted radial glial cells with impaired migration into the cerebral cortex (Pinto-Lord et al. 1982; Hunter-Schaedle 1997; Noctor et al. 1999). Both animal models have impairment in the distribution of reelin. In the case of the reeler mouse, reelin is absent. In the E24 MAM-treated ferrets, reelin is present, but the reelin-containing Cajal-Retzius cells are abnormally positioned.

4
What Prevents the Normal Laminar Pattern in E24 MAM-Treated Cortex?

Our findings led us to consider the possibility that a substance present in the normal cortex, important in maintaining the radial glia phenotype, is missing from the E24 MAM-treated cortex. We hypothesized that the cells missing as a result of MAM treatment contained, or controlled, a substance necessary for sustaining the radial nature of the radial glia. Several years ago, Hunter and Hatten (1995) proposed the idea that such a substance (or factor) existed. They found that astrocytes could be induced to elongate after exposure to embryonic cortical signals. This suggested that the ability of astrocytes to elongate or vice versa was bidirectional and dependent on the presence of diffusible cues, as opposed to a state intrinsic to the astrocyte or radial glial cell.

To test the possibility that a substance present in normal cortex is capable of sustaining the radial nature of these cells, we conducted a series of in vitro experiments that cocultured slices of E24 MAM-treated cortex with a piece of normal cortex (Fig. 4). In these studies, we cut pieces of E24 MAM-treated cortex on P0 that corresponded approximately to the parietal region. Each MAM-treated organotypic culture was either grown alone, or paired with a piece of normal cortical plate. After 2 days in culture, an injection of fluorescently labeled dextran was made into the ventricular zone of the MAM-treated slice to reveal radial glial cells. In some cocultures, immunohistochemical reactions were carried out using antibodies to reveal Cajal-Retzius cells containing reelin (the CR-50 antibody, generously provided by M. Ogawa and K. Nakajima, Tokyo, Japan) or bromodeoxyuridine (BrdU) to reveal the positions of newly generated cells.

These experiments resulted in improvement of the features previously disrupted by E24 MAM treatment. First, the morphology of the radial glia underlying the normal cortical plate explant was much improved (Fig. 5). The dextran injections revealed a relatively normal morphology of radial glial cells in the presence of P0 cortical plate from normal ferrets. In addition, the cocultures of E24 MAM-treated cortex and normal cortical plate brought about better realignment of Cajal-Retzius cells, so that they were positioned in a more normal location after coculture (Fig. 6). Finally, in addition to improving the morphology of radial glial cells and the position of Cajal-Retzius cells,

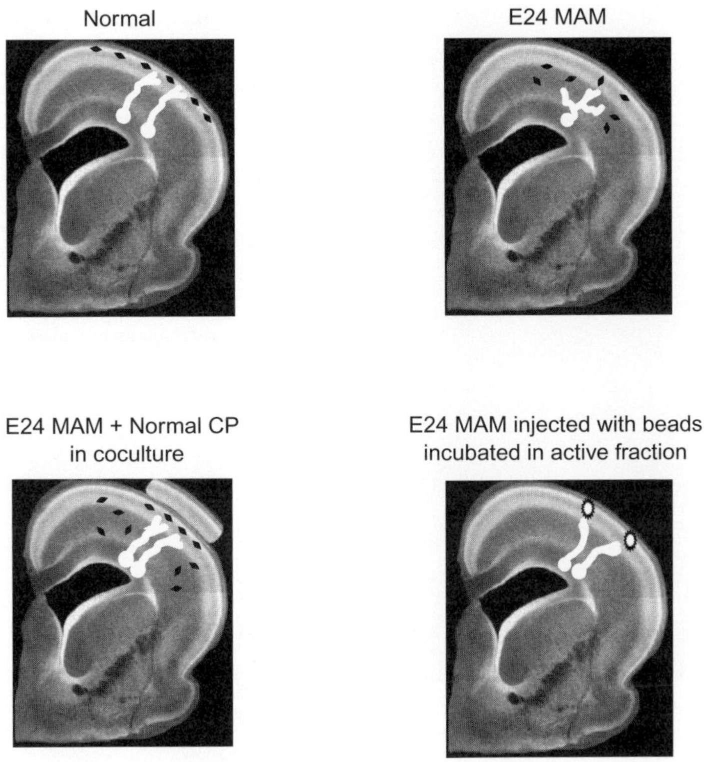

Fig. 4. Images demonstrating the experimental design to test for the presence of a radialization factor. Organotypic cultures of normal and E24 MAM-treated cortex are prepared. The *top images* demonstrate that, without any coculture, the radial glia (labeled by injection with fluorescent dextran) are relatively normal in appearance in cultures of normal brains at P0. Neurons labeled with BrdU immediately prior to preparation of cultures migrate to normal positions after 2 days in culture (indicated by *small black diamonds*). In organotypic cultures prepared from E24 MAM-treated brains, the radial glia are distorted and neurons are less likely to migrate to normal cortical positions. *Bottom left* indicates that when a slice of E24 MAM-treated brain is cocultured with a piece of normal cortical plate, the radial glia are improved in their morphology and cells labeled with BrdU migrate more effectively toward their normal positions. The image on the *bottom right* indicates that fluorescent microspheres (beads) incubated with the proper MW fraction (active fraction) of conditioned media from normal cortical plate cause radial glial cells to align toward their normal morphology

we found improved migration of newly generated cells into the cortical plate (Fig. 6).

Our findings also suggested that the dispersed Cajal-Retzius cells may follow the changes in radial glia and move into their abnormal positions along with the altered morphology of the radial fibers. If we induce early astrocytic differentiation in normal radial glia, the Cajal-Retzius cells also move into abnormal positions, suggesting that they are induced to move by the alteration in morphology of radial fibers.

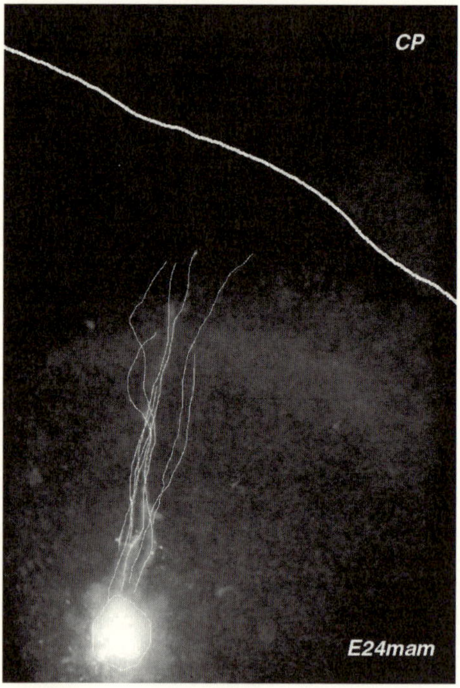

Fig. 5. Example of a coculture of an E24 MAM-treated slice of cortex paired with a piece of normal cortical plate, both obtained at P0. After growing in culture for 2 days, radial glia were revealed by an injection of fluorescent dextran in the vicinity of the ventricular zone or intermediate zone. After being exposed to normal cortical plate, the radial glial fibers resume a relatively elongated, normal morphology. The *solid line* indicates the border between the E24 MAM-treated cortical slice and the coculture of normal cortical plate

5
Is There a Radialization Factor in Normal P0 Cortex?

Our observations that coculture of normal cortical plate with E24 MAM-treated cortex improved many of the features led us to the hypothesis that a substance present in normal cortical plate causes radial glial cells to maintain their elongated morphology. We propose that the substance is present during the period of corticogenesis and migration in ferret and maintains the phenotype of extended radial glia. To further isolate the character of the substance present in normal neocortex, which improved radial glial morphology and Cajal-Retzius positioning in E24 MAM-treated cortex, we obtained conditioned media from normal slices and fractionated the media into samples of different molecular weights (MWs). We found that only media fractions of 30–50 kDa were capable of inducing radial glia to elongate in E24 MAM-treated organotypic cultures (Fig. 7). Subsequent experiments that attempted to denature the 30–50 kDa media fraction led us to believe that the substance we isolated was a protein.

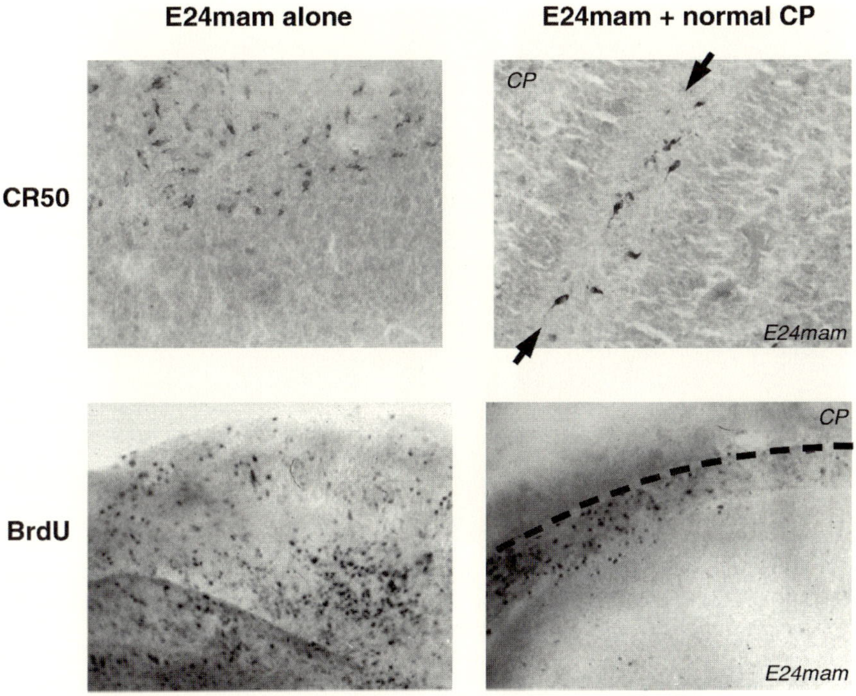

Fig. 6. *Left panels* Examples of E24 MAM-treated organotypic cultures maintained for 2 days and stained to reveal different features. *Right panels* Examples of similar E24 MAM-treated organotypic cultures paired with normal cortical plate while in culture. The *upper two panels* demonstrate CR50 (reelin) immunoreactivity and indicate that, without normal cortical plate in coculture, the Cajal-Retzius cells are scattered throughout the cerebral cortical wall in E24 MAM-treated cultures, as they are in the acute E24 MAM-treated brains (*upper left*). After coculture with normal cortical plate (*upper right*) the Cajal-Retzius cells are more normally aligned. *Arrows* indicate the border between the E24 MAM-treated slice and the normal cortical plate. *Bottom left* Example of BrdU immunoreactivity in an E24 MAM-treated slice after 2 days in culture. The BrdU was injected just prior to culture preparation. The cells are not organized in a clear laminar pattern. *Bottom right* Example of the distribution of BrdU positive cells after an E24 MAM-treated slice was placed in coculture with a piece of normal cortical plate. This results in a more precise pattern of migrating cells

6 Summary and Conclusions

Our findings suggest that a factor capable of causing the radial glial phenotype is produced by cells generated earliest during corticogenesis. Treatment with MAM on E24 in ferret prevents these cells from being present during cortical development. As a result, the radial glia are deprived of the factor, do not maintain their elongated morphology, and differentiate into astrocytes early. When we replace the missing factor, either in coculture with normal cortical plate or

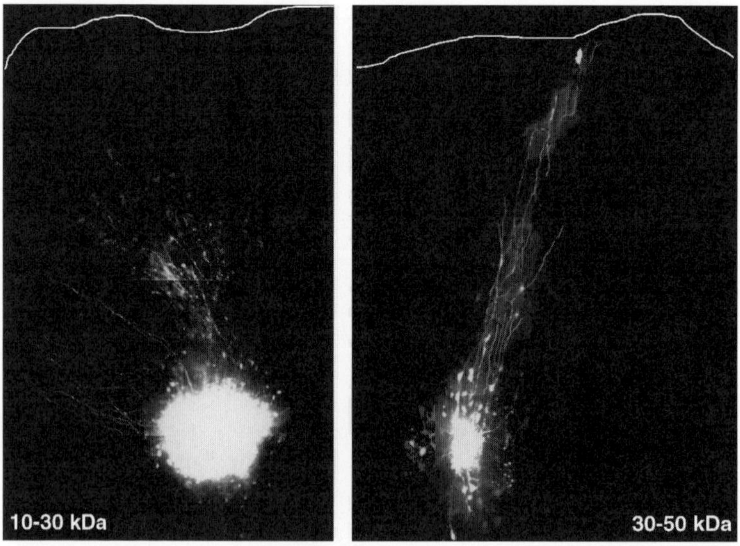

Fig. 7. Examples of E24 MAM-treated slices placed in culture with microspheres incubated in media fractions of a MW either capable of improving the morphology of radial glial cells (30–50 kDa, *right*) or not effective in altering the abnormal radial glial morphology (10–30 kDa, *left*). The radial glia are labeled by injection with fluorescent dextrans. The *solid line* indicates the pial border of the slice

by presenting the MAM-treated brain with microspheres able to slowly release the extracts of conditioned media from normal cortical plate, the radial glial fibers elongate in response. If the factor is present in the earliest generated cortical cells, there are several possible sites of production. These might include cells of the subplate or cells of layer 1. Subplate cells have been previously identified to play a supportive role for many processes in the development of the cerebral cortex and may be a likely site to produce such a factor (Shatz et al. 1992; Allendoerfer and Shatz 1994). Layer 1 cells, specifically the reelin-containing Cajal-Retzius cells, may also produce such a factor. There have been several reports strongly suggesting that Cajal-Retzius cells are important in maintaining the radial glial phenotype (Soriano et al. 1997; Super et al. 2000). Although we have not conclusively proved the existence of a radialization factor produced by early generated cortical cells, our data support the notion that such a factor exists in immature neocortex, which plays an important supportive role in normal cortical development. If this factor is withdrawn, a cascade of negative events occurs, leading to a number of cortical malformations.

References

Allendoerfer KL, Shatz CJ (1994) The subplate, a transient neocortical structure: its role in the development of connections between thalamus and cortex. Annu Rev Neurosci 17:185–218
Bentivoglio M, Mazzarello P (1999) The history of radial glia. Brain Res Bull 49:305–315
Berry M, Rogers AW (1965) The migration of neuroblasts in the developing cerebral cortex. J Anat 99:691–709
Caviness J, Sidman VS, Sidman RL (1973) Time of origin of corresponding cell classes in the cerebral cortex of normal and reeler mutant mice: an autoradiographic analysis. J Comp Neurol 148:141–152
Gilmore EC, Herrup K (2000) Cortical development: receiving reelin. Curr Biol 10(4):R162–R166
Gilmore EC, Ohshima T, Goffinet AM, Kulkarni AB, Herrup K (1998) Cyclin-dependent kinase 5-deficient mice demonstrate novel developmental arrest in cerebral cortex. J Neurosci 18:6370–6377
Gleeson JG, Walsh CA (2000) Neuronal migration disorders: from genetic diseases to developmental mechanisms. TINS 23:352–359
Hartfuss E, Galli R, Heins N, Gotz M (2001) Characterization of CNS precursor subtypes and radial glia. Dev Biol 229:15–30
Howell BW, Gertler FB, Cooper JA (1997a) Mouse disabled (mDab1): a Src binding protein implicated in neuronal development. EMBO J 16:121–132
Howell BW, Hawkes R, Soriano P, Cooper JA (1997b) Neuronal position in the developing brain is regulated by mouse disabled-1. Nature 389:733–737
Howell BW, Herrick TM, Cooper JA (1999a) Reelin-induced tryosine phosphorylation of disabled 1 during neuronal positioning. Genes Dev 13:643–648
Howell BW, Lanier LM, Frank R, Gertler FB, Cooper JA (1999b) The disabled 1 phosphotyrosine-binding domain binds to the internalization signals of transmembrane glycoproteins and to phospholipids. Mol Cell Biol 19:5179–5188
Hunter KE, Hatten ME (1995) Radial glial cell transformation to astrocytes is bidirectional: regulation by a diffusible factor in embryonic forebrain. Proc Natl Acad Sci USA 92:2061–2065
Hunter-Schaedle KE (1997) Radial glial cell development and tranformation are disturbed in reeler forebrain. J Neurobiol 33:459–472
Kwon YT, Tsai LH (2000) The role of the p35/cdk5 kinase in cortical development. Res Prob Cell Differ 30:241–253
Kwon YT, Gupta A, Zhou Y, Nikolic M, Tsai LH (2000) Regulation of N-cadherin-mediated adhesion by the p35-Cdk5 kinase. Curr Biol 10:363–372
Lambert de Rouvroit C, Goffinet AM (1998) A new view of early cortical development. Biochem Pharmacol 56:1403–1409
Levitt P, Rakic P (1980) Immunoperoxidase localization of glial fibrillary acidic protein in radial glial cells and astrocytes of the developing rhesus monkey brain. J Comp Neurol 193:815–840
Levitt P, Cooper ML, Rakic P (1981) Coexistence of neuronal and glial precursor cells in the cerebral ventricular zone of the fetal monkey: an ultrastructural immunoperoxidase analysis. J Neurosci 1:27–39
Malatesta P, Hartfuss E, Gotz M (2000) Isolation of radial glial cells by fluorescent-activated cell sorting reveals a neuronal lineage. Development 127:5253–5263
Marin-Padilla M (1998) Cajal-Retzius cells and the development of the neocortex. Trends Neurosci 21:64–71
Matsumoto H, Higa HH (1966) Studies on methylazoxymethanol, the aglycone of cycasin: methylation of nucleic acids in vitro. Biochem J 98(2):20C–22C
Meyer G, Goffinet AM, Fairen A (1999) What is a Cajal-Retzius cell? A reassessment of a classical cell type based on recent observations in the developing neocortex. Cereb Cortex 9:765–775
Miyata T, Kawaguchi A, Okano H, Ogawa M (2001) Asymmetric inheritance of radial glial fibers by cortical neurons. Neuron 31:727–741

Noctor SC, Palmer SL, Hasling T, Juliano SL (1999) Interference with the development of early generated neocortex results in disruption of radial glia and abnormal formation of neocortical layers. Cereb Cortex 9:121–136

Noctor SC, Flint AC, Weissman TA, Dammerman RS, Kriegstein AR (2001a) Neurons derived from radial glial cells establish radial units in neocortex. Nature 409:714–720

Noctor SC, Palmer SL, McLaughlin DF, Juliano SL (2001b) Disruption of layers 3 and 4 during development results in altered thalamocortical projections in ferret somatosensory cortex. J Neurosci 21:3184–3195

Palmer SL, Noctor SC, Jablonska B, Juliano SL (2001) Laminar specific alterations of thalamocortical projections in organotypic cultures following layer 4 disruption in ferret somatosensory cortex. Eur J Neurosci 13:1559–1571

Pinto-Lord MC, Evrard P, Caviness VS Jr (1982) Obstructed neuronal migration along radial glial fibers in the neocortex of the reeler mouse: a Golgi-EM analysis. Dev Brain Res 4:379–393

Rakic P (1971) Guidance of neurons migrating to the fetal monkey neocortex. Brain Res 33: 471–476

Rakic P (1972) Mode of cell migration to the superficial layers of fetal monkey neocortex. J Comp Neurol 145:61–83

Rice DS, Curran T (2001) Role of the reelin signaling pathway in central nervous system development. Annu Rev Neurosci 24:1005–1039

Schmechel DE, Rakic P (1979) A Golgi study of radial glial cells in developing monkey telencephalon: morphogenesis and transformation into astrocytes. Anat Embryol 156:115–152

Shatz CJ, Chun JJM, Luskin MB (1992) The role of the subplate in the development of the mammalian telencephalon. In: Peters A, Jones ED (eds) Cerebral cortex, vol 7. Plenum Press, New York, pp 35–56

Soriano E, Alvarado-Mallart RM, Dumesnil N, Del Rio JA, Sotelo C (1997) Cajal-Retzius cells regulate the radial glia phenotype in the adult and developing cerebellum and after granule cell migration. Neuron 18:563–577

Super H, Del Rio JA, Martinez A, Perez-Sust P, Soriano E (2000) Disruption of neuronal migration and radial glia in the developing cerebral cortex following ablation of Cajal-Retzius cells. Cereb Cortex 10:602–613

Zedeck MS, Sternberg SS, Poynter RW, McGowan J (1970) Biochemical and pathological effects of methylazoxymethanol acetate, a potent carcinogen. Cancer Res 30:801–812

Neurotrophins and Cortical Development

A. KIMBERLEY MCALLISTER[1]

Summary. The mammalian cerebral cortex requires the proper formation of exquisitely precise circuits to function correctly. These neuronal circuits are assembled during development by the formation of synaptic connections between hundreds of thousands of differentiating neurons. Although the development of the cerebral cortex has been well described anatomically, the cellular and molecular mechanisms that guide neuronal differentiation and formation of connections are just beginning to be understood. Moreover, despite evidence that coordinated patterns of activity underlie reorganization of brain circuits during critical periods of development, the molecular signals that translate activity into structural and functional changes in connections remain unknown. Recently, the neurotrophins have emerged as attractive candidates not only for regulating neuronal differentiation in the developing brain, but also for mediating activity-dependent synaptic plasticity. The neurotrophins meet many of the criteria required for molecular signals involved in neuronal differentiation and plasticity. They are present in the cerebral cortex during development and their expression is regulated by synaptic activity. In turn, the neurotrophins themselves strongly influence both short-term synaptic plasticity and long-term potentiation and depression. In addition to their functional effects, the neurotrophins also profoundly regulate the structural changes that underlie axonal and dendritic differentiation. Finally, the neurotrophins have been implicated in mediating synaptic competition required for activity-dependent plasticity during the critical period. This chapter presents and discusses the rapidly accumulating evidence that the neurotrophins are critical for neuronal differentiation and that they may be involved in activity-dependent synaptic refinement in the developing cerebral cortex.

1
Introduction

The mammalian cerebral cortex requires the proper formation of exquisitely precise neuronal circuits to function correctly. During development, these circuits are created by the formation of synaptic connections between hundreds of thousands of differentiating neurons. In order for these circuits to form correctly, neurons must elaborate axons and dendrites with specific patterns of arborization and form synapses with the correct cell types. Most synapses within the cerebral cortex are formed during a critical period of development, when immature connections are remodeled by neuronal activity and sensory experience to generate adult patterns of connectivity (reviewed in Berardi et al. 2000). Although the development of the cerebral cortex has been well

[1] Center for Neuroscience, University of California, Davis, 1544 Newton Court, Davis, California 95616, USA

Results and Problems in Cell Differentiation, Vol. 39
C.F. Hohmann (Ed.): Cortical Development

described anatomically, the cellular and molecular mechanisms that guide neuronal differentiation and formation of connections are just beginning to be understood. Furthermore, despite evidence that coordinated patterns of activity probably underlie the reorganization of brain circuits during these critical periods, the molecular signals that translate activity into structural and functional changes in connections remain unknown. Recently, the neurotrophins have emerged as attractive candidates not only for regulating neuronal differentiation in the developing brain, but also for mediating activity-dependent synaptic plasticity (reviewed in McAllister et al. 1999).

The neurotrophins are a small gene family comprised of at least four structurally and functionally related proteins – nerve growth factor (NGF), brain-derived neurotrophic factor (BDNF), neurotrophin-3 (NT-3), and neurotrophin-4/neurotrophin-5 (NT-4/5; reviewed in Chao 1992; Lindsay et al. 1994). The cellular effects of these factors are mediated through two types of receptors – a pan-neurotrophin receptor, p75, which binds all neurotrophins, and high-affinity tyrosine kinase receptors (Trk receptors) that are preferentially activated by specific neurotrophins. NGF binds to TrkA, BDNF and NT-4 bind to TrkB, and NT-3 binds to TrkC most strongly, although it can also activate TrkA and TrkB under some specific cellular contexts (reviewed in Chao 1992; Lindsay et al. 1994). At least seven *trk*B transcripts, encoding both full-length and truncated TrkB receptors (Klein et al. 1990a,b; Middlemas et al. 1991), and multiple *trk*C transcripts, encoding at least three distinct kinase domain variants and truncated forms of TrkC (Tsoulfas et al. 1993; Valenzuela et al. 1993), have been identified in the central nervous system (CNS). In general, the Trk receptors are thought to mediate directly most of the cellular effects of the neurotrophins, whereas p75 interacts with Trk receptors to modulate their signaling (reviewed in Kaplan and Miller 2000). Binding of a neurotrophin to Trk receptors induces dimerization of the receptors, phosphorylation and activation of the intracellular tyrosine kinase domain of the receptors, and initiation of a complex series of intracellular signal transduction cascades (reviewed in Kaplan and Miller 2000). In addition to local signaling, activated Trk receptors are also internalized and retrogradely transported to the nucleus, where they act to regulate gene expression (Ehlers et al. 1995; Bhattacharyya et al. 1997; Watson et al. 1999).

Historically, the neurotrophins were first described as target-derived trophic factors essential for the survival of specific neuronal populations. Indeed, it is well documented that limited availability of the neurotrophins from a synaptic target serves to match the number of afferents to the size of the target (reviewed in Levi-Montalcini 1987; Purves 1988; Barde 1989; Oppenheim 1991; Davies 1994; Snider 1994). In addition to this trophic function, the neurotrophins also have potent roles in stimulating both axon and dendrite outgrowth in many areas of the nervous system (reviewed in McAllister et al. 1999; McAllister 2001). More recently, it has been proposed that neurotrophins regulate structural and synaptic plasticity in the developing mammalian cortex by a mechanism analogous to their classic trophic function. The original neurotrophin hypothesis has been extended to include effects of the neurotrophins on individual synapses. There is increasing evidence that the neurotrophins, BDNF in particular, modulate the survival of neuronal contacts through influencing synaptic strength and/or the number of synaptic connections (reviewed in Fox and Zahs 1994; Lo 1995; Thoenen 1995; Bonhoeffer 1996; Snider and Lichtman 1996; McAllister et al. 1999). In this more modern neurotrophin hypothesis, axons compete for a limited amount of neurotrophin, secreted in an activity-dependent manner from postsynaptic sites that are coincidentally active with certain presynaptic inputs. The synapses that receive the neurotrophin are physiologically strengthened, thereby contributing to the maintenance and perhaps even proliferation of that neuronal contact. Conversely, those synapses that do not receive enough neurotrophin will be weakened and eventually eliminated (reviewed in Fox and Zahs 1994; Thoenen 1995; Lo 1995; Bonhoeffer 1996; Snider and Lichtman 1996). This chapter will present and discuss the rapidly accumulating evidence that the neurotrophins are critical for neuronal differentiation and that they may be involved in activity-dependent synaptic refinement in the developing cerebral cortex.

2 Distribution of the Neurotrophins and Their Receptors

The expression of the neurotrophins and their receptors is both regionally and developmentally regulated in the cerebral cortex. BDNF, NT-3, NT-4 and their receptors, TrkB and TrkC, are highly and specifically distributed in the cortex, whereas NGF is restricted to specific afferent projections to the cortex from sources such as the striatal and basal forebrain cholinergic neurons (reviewed in Chao 1992; Barbacid 1994; Lindsay et al. 1994). In situ hybridization studies further indicate that full-length, kinase-containing forms of TrkB and TrkC are often co-expressed on most cortical neurons, but not on non-neuronal cells. In contrast, truncated, non-catalytic forms of both TrkB and TrkC are expressed by both neurons and glial cells (reviewed in Barbacid 1994; Lindsay et al. 1994). As indicated by in situ hybridization, the expression of BDNF, NT-3, and NT-4 increases dramatically with postnatal age (Maisonpierre et al. 1990; Friedman et al. 1991; Timmusk et al. 1993). In addition, the expression of TrkB and TrkC transiently peaks between postnatal day 1 (P1) and P14, the period of maximal neuronal differentiation and synapse formation in the rodent (Ernfors et al. 1990; Dugich-Djordjevic et al. 1992; Ringstedt et al. 1993; Altar et al. 1994; Knusel et al. 1994).

Although precise localization of the neurotrophins is critical to any model for how these factors might influence neuronal differentiation, synapse formation and/or synaptic refinement, the issue of precisely where the neurotrophins and their receptors are located within the cerebral cortex is one of the most controversial aspects of the neurotrophin field. Localization of these factors is extremely difficult for several reasons. First, the neurotrophins may be expressed at very low levels in order to allow for synaptic competition for postsynaptic sites; low levels of protein are notoriously difficult to detect with any method. Second, a number of non-specific probes and antibodies have resulted in a lack of information on the expression of the multiple, complex isoforms of the Trk receptors and have led to conflicting publications within the field. In this regard, it is especially important to differentiate between full-length and truncated receptors. Although the function of truncated receptors remains unknown, they may act to decrease neurotrophin effects. Because their expression increases dramatically postnatally, any correlation between non-specific TrkB immunoreactivity and neurotrophin function could be misleading (Allendoerfer et al. 1994). Finally, since reliable, specific antibodies to the various neurotrophins have not been available until recently, most studies of the expression of these factors have used in situ hybridization to localize mRNA. Many investigators have assumed that there is a predictable relationship between message levels and levels of intracellular protein or secreted protein, but there is abundant evidence within the neurotrophin field that no consistently clear relationship exists.

Despite these historical problems, the expression of the neurotrophins and their receptors is best studied in the visual system, including the primary visual cortex. In general, TrkB expression increases specifically at times of neuronal growth and synaptogenesis in separate locations in the developing visual system (Cabelli et al. 1996). In the primary visual cortex, BDNF mRNA is found in neurons in layers 5 and 6 prior to eye-opening; this expression spreads to the superficial cortical layers shortly after eye-opening (Lein and Shatz 2000). Within layer 4, BDNF mRNA is low initially and rises to high levels by the end of the critical period for ocular dominance column formation (Lein and Shatz 2000). Although the expression patterns of NT-3 and TrkC are not as well characterized, current data is consistent with the view that they are found at high levels within the cortical layers early in development, but disappear in the adult (Lein and Shatz 2000). Despite this detailed information for the visual system, it is important to acknowledge that the expression of the neurotrophins and their receptors is likely to differ between species and cortical areas.

Regardless of the precise laminar localization of the neurotrophins and their receptors, it appears that at least some of the neurotrophins and their receptors are found specifically at synapses in the developing cortex. BDNF has been localized to synaptic junctions of cultured cortical neurons using a number of approaches (Haubensak et al. 1998; Kryl et al. 1999; Aoki et al. 2000). TrkB is also located at synapses both pre- and postsynaptically in the CNS. TrkB was first demonstrated to be an intrinsic component of the postsynaptic density using biochemical analysis (Wu et al. 1996). The full-length form of TrkB has been further localized to axons, axon terminals, and dendritic spines of neurons in the hippocampus and cerebral cortex, using immunocytochemistry (Drake et al. 1999; Aoki et al. 2000). TrkB is expressed primarily by excitatory, glutamatergic neurons and, to a lesser extent, by inhibitory neurons. Thus, TrkB is found both pre- and postsynaptically, supporting multiple roles for this neurotrophin in the cerebral cortex.

2.1 Regulation of the Neurotrophins by Activity

One of the most interesting aspects of neurotrophin expression is its regulation by electrical and synaptic activity. This activity-dependent regulation was first demonstrated in the hippocampus, where high basal levels of the neurotrophins are present. NGF and BDNF mRNA levels are rapidly and potently increased by epileptiform activity in the hippocampus and cerebral cortex (Gall and Isackson 1989; Zafra et al. 1990; Ernfors et al. 1991; Isackson et al. 1991; Dugich-Djordjevic et al. 1992). BDNF mRNA expression is dramatically affected, increasing by over 6-fold in the dentate gyrus within 30 min after seizure induction (Ernfors et al. 1991). In some cases, similar increases in electrical activity reduce NT-3 expression (Castren et al. 1993; Elmer et al. 1996; Mudo et al. 1996).

Manipulations of more specific forms of neuronal activity have been demonstrated to regulate neurotrophin mRNA levels in neuronal cultures. Depolarization of cultured hippocampal neurons with high potassium or glutamate receptor agonists potently increases BDNF and NGF mRNAs (Zafra et al. 1990, 1991, 1992; Lu et al. 1991; Berzaghi et al. 1993; Berninger et al. 1995; Lindholm et al. 1994; Lauterborn et al. 2000). Conversely, neurotrophin levels are decreased in the presence of the inhibitory neurotransmitter γ-aminobutyric acid (GABA) (Zafra et al. 1991, 1992; Berzaghi et al. 1993; Berninger et al. 1995). Most of these effects of activity on neurotrophin and Trk receptor expression are rapid; for example, following treatment with the non-NMDA receptor agonist, kainate, BDNF mRNA levels increase by greater than 10-fold within 3 h after stimulation (Zafra et al. 1991).

More physiological levels of synaptic activity also regulate neurotrophin and Trk receptor expression. Induction of long-term potentiation (LTP) in slices, and contextual learning in vivo, rapidly and selectively enhance BDNF expression in the hippocampus, with little or no effect on the other neurotrophins (Patterson et al. 1992; Castren et al. 1993; Dragunow et al. 1993; Hall et al. 2000). Changes in the levels of even more physiological stimuli, such as light, also regulate BDNF and TrkB levels in the visual cortex (Castren et al. 1992; Bozzi et al. 1995; Schoups et al. 1995; Rocamora et al. 1996; Capsoni et al. 1999; Lein and Shatz 2000). In fact, BDNF mRNA levels in the visual cortex are an accurate molecular indicator of the local activity levels of cortical neurons (Lein and Shatz 2000). All of these results suggest that the neurotrophins, BDNF in particular, can be upregulated by enhanced physiological activity. However, it will be important in the future to design experiments that allow more subtle and local manipulations of neuronal activity at individual synapses. These studies do show that neurotrophin mRNA levels can be upregulated by depolarization, but whether normal levels of synaptic activity cause similar changes at single synapses remains unknown.

In addition to regulation of overall neurotrophin and Trk receptor mRNA levels, neuronal activity may also recruit BDNF and TrkB mRNAs to dendrites and enhance TrkB receptor insertion into the cell membrane. First, depolarization of cultured hippocampal neurons induces targeting of BDNF and TrkB mRNAs to dendrites (Tongiorgi et al. 1997). This activity-dependent mRNA targeting requires calcium influx and a PI-3 kinase pathway (Righi et al. 2000). Second, depolarization with high potassium, or cAMP application, causes a nearly 4-fold increase in surface TrkB levels in cultured retinal ganglion neurons (Meyer-Franke et al. 1998). Similarly, high-frequency stimulation also recruits TrkB receptors to the neuronal membrane from intracellular pools in cultured hippocampal neurons (Du et al. 2000). As this membrane recruitment does not require new protein synthesis, these results suggest that synaptic activity rapidly recruits BDNF and TrkB to dendritic locations from intracellular stores. Although the effects of activation of single synapses have not yet been investigated, activity-dependent recruitment of TrkB and BDNF to dendrites may provide a mechanism for selectively and locally increasing these factors at sites of high-frequency synaptic transmission.

2.2 Effects of Activity on Neurotrophin Secretion

In addition to activity-dependent regulation of overall neurotrophin expression and dendritic localization of protein and mRNA, neurotrophin secretion may also be tightly regulated by synaptic activity. If the neurotrophins function as retrograde signals that act to selectively strengthen coincidentally active synapses, then release should also occur only at certain sites along an activated neuron's dendritic tree closely apposed to active presynaptic terminals.

Although secretion of neurotrophins in the peripheral nervous system (PNS) is mostly constitutive (Barth et al. 1984), recent evidence suggests that cortical neurons may release neurotrophins in an activity-regulated manner (Blochl and Thoenen 1995). Extensive depolarization caused by high potassium, glutamate, and acetylcholine stimulates secretion of NGF and BDNF from hippocampal slices and dissociated hippocampal neurons overexpressing NGF (Blochl and Thoenen 1995, 1996; Goodman et al. 1996). Furthermore, endogenous BDNF has been localized to dense-core vesicles in hippocampal neuron dendrites (Fawcett et al. 1997; Smith et al. 1997). Using an assay for neurotrophin effects on synaptic strength, Poo and coworkers demonstrated a synapse-specific effect of retrogradely secreted NT-4 from postsynaptic muscle cells on presynaptic function (Wang and Poo 1997). In addition to evidence for postsynaptic neurotrophin secretion, followed by binding and activation of presynaptic TrkB receptors, there is increasing evidence that the neurotrophins are also secreted from axon terminals to activate postsynaptic TrkB receptors (von Bartheld et al. 1996; Aloyz et al. 1999; von Bartheld and Butowt 2000). Most compelling, activity-dependent transfer of BDNF, fused to the green fluorescent protein (GFP), from presynaptic axons to postsynaptic neurons has been visualized directly in cultured hippocampal neurons (Kohara et al. 2001).

3 Regulation of Synaptic Plasticity by the Neurotrophins

3.1 Acute Effects on Synaptic Function

One of the most potent and rapid effects of the neurotrophins is to modulate synaptic activity. This aspect of neurotrophin function is one of the primary reasons for the excitement surrounding the possibility that these factors are critical for activity-dependent plasticity during cortical development. It is hypothesized that the neurotrophins mediate activity-dependent synaptic refinement by controlling simultaneously the physiological strength and the morphology of neuronal connections. The first evidence for effects of the neurotrophins on synaptic transmission was provided by acute application of BDNF and NT-3 at developing *Xenopus* neuromuscular synapses in culture. BDNF and NT-3, but not NGF, increase the frequency but not the amplitude of spontaneous synaptic currents (mEPSCs) and the amplitude of nerve-evoked excitatory postsynaptic currents (EPSCs) in these cultures (Lohof et al. 1993; Wang et al. 1995). These effects are rapid (occurring within 5 min), large (>6-fold; Lohof et al. 1993), depend on the continued presence of neurotrophin and presynaptic depolarization, and are gated by cAMP (Boulanger and Poo 1999a,b). Recently, a more direct link between postsynaptic neurotrophin secretion at synapses and increased neurotransmitter release has been demonstrated. In that report, activity-dependent secretion of NT-4 from postsynaptic myocytes overexpressing NT-4 dramatically enhances presynaptic acetylcholine release (Wang and Poo 1997).

Neurotrophins also potently modulate synaptic transmission between cortical neurons. At central synapses, neurotrophins enhance excitatory synaptic transmission and suppress inhibitory transmission (reviewed in Poo 2001). BDNF and NT-4 rapidly enhance spontaneous synaptic activity in cultured hippocampal neurons by increasing the frequency, but not the amplitude, of mEPSCs, implying a presynaptic locus of action (Lessmann et al. 1994; Li et al. 1998b). In young rat cortical slices, acute BDNF or NGF application also increases the amplitude of EPSCs and the frequency, but not the amplitude, of mEPSCs (Carmignoto et al. 1997). Similarly, NT-3 also potentiates neuronal activity in cultured cortical neurons (Kim et al. 1994). Acute application of BDNF or NT-3, but not NGF, to hippocampal slices potentiates synaptic transmission at the Schaffer collateral-CA1 synapse that does not occlude LTP (Kang and Schuman 1995, 1996). This BDNF-induced potentiation of synaptic transmission is due to a presynaptic action of BDNF, as demonstrated by an elegant experiment in which TrkB receptors were selectively blocked either pre- or postsynaptically using a dominant negative approach (Li et al. 1998a). Also consistent with a presynaptic site of action, BDNF increases levels of many synaptic vesicle proteins (Takei et al. 1997) and promotes phosphorylation of synapsin 1 (Jovanovic et al. 1996).

The ability of the neurotrophins to acutely enhance neurotransmitter release is consistent with the classic model of the neurotrophins as retrograde signaling molecules; however, neurotrophins also potently modulate postsynaptic aspects of synaptic transmission and, thus, may also act as anterograde signals. Physiologically, BDNF can increase the magnitude of EPSCs in hippocampal cultures, without affecting mEPSC frequency (Levine et al. 1995; Sherwood and Lo 1999; Bolton et al. 2000). In addition, the neurotrophins exert fast effects on other NMDA channel properties (Jarvis et al. 1997; Crozier et al. 1999) and BDNF, in particular, can lead to phosphorylation of the NR1 NMDA receptor subunit within minutes of application (Suen et al. 1997). These postsynaptic effects of the neurotrophins may be mediated by calcium elevation in dendrites (Berninger et al. 1993; Stoop and Poo 1996; Canossa et al. 1997; Jarvis et al. 1997). Perhaps most intriguing, BDNF and NT-4 were recently shown to act like neurotransmitters at cortical synapses, causing membrane depolarization within a few milliseconds through a proposed interaction between activated TrkB and a new form of sodium channel (Kafitz et al. 1999).

In addition to these rapid effects on excitatory synaptic transmission, the neurotrophins also potently regulate inhibitory transmission. BDNF increases neuropeptide expression in hippocampal interneurons (Marty et al. 1996a,b) and enhances GABA uptake and cell body size in embryonic cortical neurons (Widmer and Hefti 1994). BDNF and NT-3 also acutely inhibit GABAergic synaptic transmission between hippocampal and cortical neurons (Kim et al. 1994; Tanaka et al. 1997; Frerking et al. 1998; but see Bolton et al. 2000), and influence activity-dependent regulation of inhibition in neocortical cultures (Rutherford et al. 1997). Finally, BDNF was recently shown to be critical for the maturation of inhibitory circuitry (Huang et al. 1999).

3.2 Long-Term Potentiation and Depression

During experience-dependent development of many areas of the mammalian cerebral cortex, synapses are strengthened by correlated activity and are actively weakened by non-correlated activity (reviewed in Katz and Shatz 1996). Long-term potentiation (LTP) and depression (LTD) are the most convincing cellular models to date for synaptic strengthening and weakening in the developing cortex (Bear and Rittenhouse 1999; Berardi et al. 2000). Current models for cortical development also suggest that synapses are ultimately strengthened or weakened based on their ability to obtain a limited target-derived signal (reviewed in Snider and Lichtman 1996). Because of their activity-dependent upregulation of expression, their activity-dependent release from postsynaptic neurons, and their rapid enhancement of synaptic function, neurotrophins are attractive candidates for a retrograde factor that strengthens synapses during activity-dependent cortical development.

In support of a role for BDNF in LTP, the ability to induce LTP at the Schaffer collateral-CA1 synapse in the hippocampus is dramatically reduced in BDNF knockout mice (Korte et al. 1995; Patterson et al. 1996; Xu et al. 2000a). This defect in LTP induction is likely due to a lack of BDNF, rather than to a non-specific developmental defect. The deficit in LTP can be rescued either by exogenous application of BDNF (Patterson et al. 1996) or by transfecting hippocampal slices with a BDNF-containing adenovirus (Korte et al. 1996). Acute inhibition of BDNF with function-blocking antibodies or Trk-IgGs also prevents LTP induction in adult hippocampal and young cortical slices (Figurov et al. 1996; Akaneya et al. 1997; Kang et al. 1997; Chen et al. 1999) and exogenous BDNF enhances the magnitude of LTP in the visual cortex (Akaneya et al. 1997). LTD also appears to be influenced by neurotrophins in the visual cortex. Pretreatment of visual cortical slices with BDNF, but not NGF, prevents LTD of field responses from being induced by low-frequency stimulation (Akaneya et al. 1996; Huber et al. 1998).

Neurotrophins released during LTP may also stimulate the protein synthesis required for the late phase of LTP (Frey et al. 1988; Nguyen et al. 1994; Kang et al. 1997). The induction and maintenance of the late phase of LTP requires BDNF (Kang et al. 1997) and neurotrophin-induced LTP requires local dendritic protein synthesis (Kang and Schuman 1996). Thus, LTP may stimulate secretion of the neurotrophins, which, in turn, stimulate local protein synthesis, thereby exerting local control over the protein complement of individual synapses (Crino and Eberwine 1996).

Although there is substantial controversy regarding the locus of alterations involved in LTP and LTD in the cortex, neurotrophins (especially BDNF) appear to play a significant role. Whether neurotrophins act as a permissive signal enabling synaptic machinery to be maintained at a level required for induction of plastic changes, or whether they function as genuinely instructive factors, whose release is regulated sufficiently tightly to confer some degree of synapse specificity, remains unclear. The same issues complicate interpretation of neurotrophin effects in developing systems as well, such as the ocular dominance column system in the visual cortex (see below).

4
Neurotrophins and Structural Synaptic Plasticity

Cortical development not only requires the precise differentiation of neurons within the six cortical layers, but also requires activity-dependent rearrangements of connections within many cortical areas (reviewed in Katz and Shatz 1996). The neurotrophins are believed to be critical for both processes. For example, the rearrangements that result in eye-specific layers in the lateral geniculate nucleus (LGN) and ocular dominance columns in the visual cortex involve selective elaboration of coincidentally active connections in one layer or column and pruning of inappropriate connections in other layers or columns (Hubel and Wiesel 1970; Shatz 1983; reviewed in Katz and Shatz 1996; but see Crowley and Katz 2000). Molecular signals that mediate these synaptic rearrangements should be capable of simultaneously modifying the physiological properties of neurons and the structure of axonal or dendritic arborizations; the neurotrophins are one of the only molecular signals known to date that can subserve both of these functions (reviewed in McAllister et al. 1999).

4.1 Axonal Growth

The neurotrophins are potent regulators of axonal growth in both the peripheral and central nervous systems. In fact, NGF was first purified based on its capacity to stimulate neurite growth. Addition of an NGF-secreting sarcoma to a chick sensory ganglion explant stimulates rapid growth of a dramatic halo of neurites in only a few hours (Levi-Montalcini et al. 1954). Since that first report, each of the neurotrophins has been shown to stimulate neurite outgrowth of specific populations of neurons in the PNS (reviewed in Eide et al. 1993; Snider 1994). In the last few years, the ability of the neurotrophins to influence axon guidance in the CNS has begun to be explored for the first time.

A number of early in vitro experiments suggested that the neurotrophins may act as chemotropic axon guidance molecules. Diffusible gradients of neurotrophins potently regulate chemotropic axon guidance of several peripheral neuronal populations in vitro (Gunderson and Barrett 1979). Interestingly, these effects on axon guidance can be either attractive or repulsive depending on the neuronal and neurotrophin type (Ming et al. 1997; Paves and Saarma 1997; Song et al. 1997). However, despite the compelling nature of the in vitro data, evidence is accumulating against a chemotropic role for the neurotrophins in vivo. Most importantly, the phenotypes of the neurotrophin knockout mice do not show the clear deficits in axon pathfinding that are expected for true chemotropic molecules (reviewed in Snider 1994).

Independent of their controversial effects on axon guidance, it is clear that the neurotrophins potently influence the complexity of axonal arbors in the CNS. For example, the complexity of retinal axons is dramatically enhanced by BDNF, NT-3, and NT-4 (but not NGF) when retinas are cocultured with chick optic tecta (Inoue and Sanes 1997). The branching and complexity of retinal ganglion cell axons is also increased within 2 h by in vivo infusion of BDNF, but not other neurotrophins, into the optic tectum of *Xenopus* tadpoles (Cohen-Cory and Fraser 1995). Conversely, axon growth and complexity is reduced after injection of function-blocking antibodies (Cohen-Cory and Fraser 1995).

Like other axon guidance molecules, the effects of the neurotrophins on axonal growth are complex and specific for particular cell types. In addition to their growth-promoting effects, the neurotrophins can also decrease axon growth and branching depending on the cellular context. For example, BDNF rapidly induces axonal growth cone collapse and retraction of cultured embryonic *Xenopus* spinal neurons (Wang and Zheng 1998). Furthermore, in the developing visual cortex, the neurotrophins may act in a bifunctional manner to guide the complex choices that underlie development of vertical connections between cortical layers. Specifically, NT-3 is attractive for axons from neurons in layer 6, while it is repulsive for axons from neurons in layers 2/3 in cultured cortical slices (Castellani and Bolz 1999).

The ability of the neurotrophins to influence axon growth in a bifunctional manner may be determined by intracellular signal transduction cascades. Poo and colleagues have recently shown that the levels of cyclic nucleotides can change the response of an axonal growth cone to a single neurotrophin from attractive to repulsive (Song et al. 1997). In fact, cAMP or cGMP levels can influence the effects of several axon guidance molecules in vitro (Song et al. 1997). Thus, it is possible that an axonal growth cone could respond to the same neurotrophin with opposite cellular responses at different times based on modulation of intracellular cAMP levels by other environmental factors. This is an intriguing model but it needs further characterization; these manipulations of cyclic nucleotides are large-scale, non-physiological manipulations and there is not yet any evidence for such a function for cyclic nucleotides in vivo.

The bifunctional effects of the neurotrophins on axon guidance could also be mediated by the Rho GTPase family (MacKay and Hall 1998). Recent evidence suggests that the neurite-promoting effects of the neurotrophins may be mediated by a direct physical interaction between the neurotrophin receptor, p75, and the RhoA GTPase (Yamashita et al. 1999). This interaction provides a potential link between the neurotrophins and the actin cytoskeleton, as the Rho GTPases control the organization of actin in axonal growth cones (MacKay and Hall 1998). Specifically, the ability of NGF to enhance the growth of ciliary neurons through p75 requires inactivation of RhoA (Yamashita et al. 1999). These results suggest that the cytoskeleton of axonal growth cones might be remodeled rapidly in response to neurotrophins encountered during axon pathfinding. Further clarification of the roles of the neurotrophin receptors, including p75, and the signal transduction cascades that ultimately link neurotrophin binding to changes in the actin cytoskeleton will increase our understanding of the influence of the neurotrophins on axonal growth in the future.

4.2 Dendritic Growth

Because dendritic arborizations determine and restrict the number, pattern, and types of synapses received by each neuron, dendritic growth is critical for the proper functioning of the brain, and especially the cerebral cortex. Neurotrophins play a central role in regulating the complexity and pattern of dendritic growth in both the peripheral and central nervous systems (reviewed in McAllister 2000). Neurotrophins were first demonstrated to regulate dendritic growth in the PNS. Systemic treatment of neonatal and adult rats with NGF for 1–2 weeks increased dendritic complexity of sympathetic ganglion cells, whereas injections of NGF antiserum prevented dendritic growth (Snider 1988; Ruit et al. 1990; Ruit and Snider 1991).

In the CNS, the neurotrophins also regulate dendritic growth of many types of neurons in many different brain regions (reviewed in McAllister et al. 1999; McAllister 2000). In particular, the neurotrophins potently regulate dendritic growth of pyramidal neurons in the developing visual cortex (McAllister et al. 1995; Baker et al. 1998). Each of the four neurotrophins, when applied to organotypic cortical slices for only 36h, rapidly enhances dendritic growth and complexity of cortical pyramidal neurons (McAllister et al. 1995). Neurons in each of the six cortical layers respond to a specific subset of neurotrophins and, within a single cortical layer, each neurotrophin elicits a unique pattern of changes in dendritic morphology (McAllister et al. 1995). Moreover, cortical pyramidal neurons that overexpress BDNF exhibit enhanced dendritic dynamics, a reduced number of dendritic spines, and an increased number of primary dendrites (Horch et al. 1999). Thus, increased BDNF may induce structural instability in both dendrites and spines. The wide range of neurotrophin effects on dendrite growth, in addition to their laminar specificity, implies that these factors may act instructively to regulate the development of specific patterns of dendritic arborizations in the cerebral cortex.

Most importantly, endogenous neurotrophins also influence dendritic growth. By blocking endogenous factor with Trk receptor-bodies, McAllister and colleagues clearly demonstrated that endogenous neurotrophins powerfully influence the complexity of dendritic arbors of cortical pyramidal neurons (McAllister et al. 1997). Consistent with results from adding exogenous neurotrophin (McAllister et al. 1995), endogenous BDNF is required for growth and maintenance of dendritic arbors of layer 4 neurons, while endogenous NT-3 is required for growth and maintenance of dendritic arbors of layer 6 neurons in the developing visual cortex (McAllister et al. 1997).

Blocking endogenous neurotrophins revealed a surprising and more complex level of regulation of dendritic growth. Endogenous neurotrophins were found not only to enhance dendritic growth as expected, but also to limit growth and even cause dendritic retraction, depending on the layer-specific location of the neurons examined (McAllister et al. 1997). Moreover, the endogenous TrkB ligands and NT-3 oppose one another in regulating dendritic growth. In layer 4, NT-3 limits dendritic growth caused by BDNF while, in layer 6, BDNF inhibits dendritic growth stimulated by NT-3 (McAllister et al. 1997; but see Yacoubian and Lo 2000). Thus, neurotrophins have bifunctional effects on dendritic growth in addition to their bifunctional effects on axon guidance. These antagonistic roles for BDNF and NT-3 provide a potential mechanism by which dendritic growth and retraction could be locally and dynamically regulated by intercellular interactions.

Endogenous neurotrophins also play important roles in maintaining dendritic arborizations of cortical pyramidal neurons at later stages of development. Using a targeted gene knockout approach in which TrkB was deleted from primarily cortical pyramidal neurons, endogenous BDNF was shown to be critical for maintenance of dendrites of cortical neurons after they had completed their initial formation (Xu et al. 2000b). These mice, which lose TrkB expression in the fourth postnatal week, exhibit massive neuronal death and dendritic retraction of cortical pyramidal neurons. Importantly, these results implicate neurotrophins in the maintenance of cortical dendritic arbors and suggest a potential role for neurotrophins in the etiology of neurodegenerative diseases (Xu et al. 2000b).

Although the neurotrophins certainly regulate the growth of dendrites in the developing cerebral cortex, there are many other molecular signals that also influence dendritic growth, including neuronal activity (reviewed in McAllister 2000). In fact, recently it has been proposed that the neurotrophins act as molecular mediators of the effects of neuronal activity on many cellular processes during development. In support of this hypothesis, the neurotrophins influence dendritic growth of preferentially active neurons (McAllister et al. 1996). Inhibition of action potentials, glutamatergic synaptic transmission, or L-type calcium channels, each completely prevents the robust enhancement of dendritic growth caused by BDNF. Thus, neurons must be active in order to respond to the growth-promoting effects of BDNF (McAllister et al. 1996). This requirement for simultaneous synaptic activity and neurotrophin signaling provides a possible mechanism for selectively enhancing growth of dendrites receiving inputs from active neurons in the developing cerebral cortex.

4.3 Synapse Formation and Maintenance

In addition to influencing the growth and complexity of axons and dendrites, it has been proposed that the neurotrophins also control the overall number of synapses in a particular brain region. Although there is not yet any direct evidence that the neurotrophins control synapse formation or maturation in the CNS, there are an increasing number of reports consistent with such a role (Wang et al. 1995; Causing et al. 1997; Loeb and Fischbach 1997; Martinez et al. 1998; Vicario-Abejon et al. 1998; Gonzalez et al. 1999; Pozzo-Miller et al. 1999; Wells et al. 1999; Collin et al. 2001). For example, in the superior cervical ganglia of transgenic mice over-expressing BDNF, synaptic innervation density is enhanced 2.5-fold and the density of synapses is decreased in BDNF knockout mice (Causing et al. 1997). These BDNF knockout mice also exhibit structural defects in their presynaptic terminals, including a decreased number of docked synaptic vesicles and reduced expression of several synaptic vesicle proteins (Pozzo-Miller et al. 1999). TrkB and TrkC knockout mice also have a decreased synaptic density and less complex axonal arborizations (Martinez et al. 1998).

The neurotrophins have also been implicated in synapse formation in the developing CNS. BDNF and NT-3 contribute to the development of presynaptic differentiation and the formation of functional excitatory and inhibitory synapses in E16 dissociated hippocampal cultures (Vicario-Abejon et al. 1998; Collin et al. 2001). The TrkB ligands are also required for activity-dependent inhibitory synapse formation in organotypic cerebellar cultures (Seil and Drake-Baumann 2000). Unfortunately, despite the increasing number of reports of neurotrophin effects on synapse number, it is almost impossible to attribute the effects of the neurotrophins directly to regulating synapse number; these results are often complicated by additional neurotrophin effects on survival or growth of axons or dendrites. Most of the effects of the neurotrophins on synapse number reported to date could result from direct effects of this factor on neuronal survival or structure that indirectly influence the number of synaptic connections.

4.4 Activity-Dependent Plasticity

Perhaps the most exciting function for the neurotrophins is the possibility that they might mediate activity-dependent refinement of connections during cortical development (reviewed in Katz and Shatz 1996; McAllister et al. 1999). The influence of neural activity on the formation of cortical circuits has been most extensively studied in the visual system. Activity-driven structural changes form eye-specific layers in the LGN and ocular dominance columns in visual cortex (reviewed in Shatz 1990; Katz and Shatz 1996; but see Crowley and Katz 2000). Although the spectrum of anatomical and physiological changes that occur in response to synaptic activity during development are well characterized, the cellular and molecular mechanisms by which activity is translated into these changes in circuitry remain poorly understood.

The neurotrophins are attractive candidates for molecular mediators of the effects of activity in influencing neuronal connectivity in the developing cortex for several reasons. First, the neurotrophins and their receptors are expressed in the cerebral cortex during the period of activity-dependent synaptic refinement (described above; Cabelli et al. 1996; Lein and Shatz 2000). Second, neurotrophin and Trk receptor expression can be rapidly controlled by synaptic activity within the visual system, especially during the time of ocular dominance column formation (described above; Castren et al. 1992; Schoups et al. 1995; Capsoni et al. 1999; Rossi et al. 1999; Lein and Shatz 2000). Third, both acute synaptic transmission and long-term synaptic efficacy in the developing visual cortex are rapidly and potently regulated by the neurotrophins (described above; Kim et al. 1994; Akaneya et al. 1996, 1997; Carmignoto et al. 1997; Takei et al. 1997; Huber et al. 1998; Sala et al. 1998; Kafitz et al. 1999). The synaptic scaling that results from short-term, global activity manipulations may also be controlled by BDNF (Rutherford et al. 1997; Desai et al. 1999). Fourth, the neurotrophins also regulate axonal and dendritic growth in the developing visual cortex in addition to their effects on synaptic transmission (as described above). Thus, the neurotrophins meet all of the criteria required for a potential molecular mediator of the effects of activity in refining connections within the visual system.

Manipulation of neurotrophin levels in vivo also supports their potential function as molecular mediators of the activity-dependent synaptic refinement required for ocular dominance column formation. The first indication that the neurotrophins might be involved in activity-dependent synaptic plasticity in the visual cortex came from a series of experiments performed by Maffei and colleagues. They demonstrated that infusion of NGF, or NGF antibodies, into the cerebral ventricles of young rats prevents the shift in ocular dominance caused by monocular deprivation (Maffei et al. 1992; Berardi et al. 1993; Carmignoto et al. 1993; Domenici et al. 1993, 1994). These findings led to the hypothesis that LGN axons normally participate in activity-dependent competition for a limiting supply of neurotrophin (NGF in their experiments) secreted by neurons in layer 4 of the visual cortex. More recent experiments support this general hypothesis, but suggest that it may be a TrkB ligand, BDNF or NT-4, that is the focus of LGN axon competition (Riddle et al. 1995; Hata et al. 2000; Lodovichi et al. 2000; but see Gu et al. 1994; Galuske et al. 1996, 2000). For example, both addition of excess BDNF or NT-4 and removal of endogenous BDNF and NT-4 in vivo prevent ocular dominance columns from forming; NGF and NT-3 have no effect in either manipulation (Cabelli et al. 1995, 1997). Thus, there is an emerging consensus that neurotrophins are important for activity-dependent plasticity in the developing visual cortex at some level. However, there is considerable controversy about the synapse specificity and cellular mechanisms of the neurotrophin effects (reviewed in McAllister et al. 1999).

5
Concluding Remarks

The neurotrophins play many roles in the development of the cerebral cortex; the clearest functions for the neurotrophins during development are to modulate synaptic transmission, control long-term synaptic strength, and influence axonal and dendritic growth. However, direct unequivocal evidence for roles of the neurotrophins in mediating synapse formation and activity-dependent plasticity is still lacking. In particular, the cellular and molecular mechanisms by which the neurotrophins influence activity-dependent synaptic competition remain elusive. Much of the controversy surrounding the role of neurotrophins in activity-dependent development and plasticity stems from inherent difficulties in interpreting results from experiments published to date, due mostly to ambiguities resulting from global manipulation of the neurotrophins in intact tissue. Several critical issues will have to be resolved before conclusions about these latter neurotrophin functions can be solidified. First, more thorough studies of the subcellular distribution of the neurotrophins and Trk receptors, as well as the locations and mechanisms of their secretion, are needed to advance our understanding of neurotrophin function. Second, the specific effects of the neurotrophins, and their interactions with synaptic activity, at individual synapses must be clarified. This issue is especially important in determining whether neurotrophins act through synapse-specific mechanisms, a requirement for an instructive function in activity-dependent plasticity. Finally, it is also crucial to determine if neurotrophins increase synaptic strength exclusively of inputs coincidentally active with the postsynaptic neuron under physiological conditions, as required in most Hebbian models of synaptic competition (Hebb 1949). It is exciting that we now have a strong candidate molecular signal for mediating synaptic plasticity, but the truly definitive experiments proving such a function still remain to be performed.

References

Akaneya T, Tsumoto T, Hatanaka H (1996) Brain-derived neurotrophic factor blocks long-term depression in rat visual cortex. J Neurophysiol 76:4198–4201

Akaneya Y, Tsumoto T, Kinoshita S, Hatanaka H (1997) Brain-derived neurotrophic factor enhances long-term potentiation in rat visual cortex. J Neurosci 17:6707–6716

Allendoerfer KL, Cabelli RJ, Escandon E, Kaplan DR, Nikolics K, Shatz CJ (1994) Regulation of neurotrophin receptors during the maturation of the mammalian visual system. J Neurosci 14:1795–1811

Aloyz R, Fawcett JP, Kaplan DR, Murphy RA, Miller FD (1999) Activity-dependent activation of TrkB neurotrophin receptors in the adult CNS. Learn Mem 6:216–231

Altar CA, Siuciak JA, Wright P, Ip NY, Lindsay RM, Wiegand SJ (1994) In situ hybridization of trkB and trkC receptor mRNA in rat forebrain and association with high-affinity binding of $^{[125]}$BDNF, $^{[125]}$NT-4/5 and $^{[125]}$NT-3. Eur J Neurosci 6:1389–1405

Altar CA, Cai N, Bliven T, Juhasz M, Conner JM, Acheson AL, Lindsay RM, Wiegand SJ (1997) Anterograde transport of brain-derived neurotrophic factor and its role in the brain. Nature 389:856–860

Aoki C, Wu K, Elste A, Len G-W, Lin S-Y, McAuliffe G, Black I (2000) Localization of brain-derived neurotrophic factor and TrkB receptors to postsynaptic densities of adult rat cerebral cortex. J Neurosci Res 59:454–463

Baker RE, Dijkhuizen PA, Van Pelt J, Verhaagen J (1998) Growth of pyramidal, but not non-pyramidal, dendrites in long-term organotypic explants of neonatal rat neocortex chronically exposed to neurotrophin-3. Eur J Neurosci 10:1037–1044

Barbacid M (1994) The Trk family of neurotrophin receptors. J Neurobiol 25:1386–1403

Barde YA (1989) Trophic factors and neuronal survival. Neuron 2:1525–1534

Barth EM, Korsching S, Thoenen H (1984) Regulation of nerve growth factor synthesis and release in organ cultures of rat iris. J Cell Biol 99:839–843

Bear MF, Rittenhouse CD (1999) Molecular basis for induction of ocular dominance plasticity. J Neurobiol 41:83–91

Berardi N, Domenici L, Parisi V, Pizzorusso T, Cellerino A, Maffei L (1993) Monocular deprivation effects in the rat visual cortex and lateral geniculate nucleus are prevented by nerve growth factor (NGF). I. Visual cortex. Proc R Soc London (Biol) 251:17–23

Berardi N, Pizzorusso T, Maffei L (2000) Critical periods during sensory development. Curr Opin Neurobiol 10:138–145

Berninger B, Garcia DE, Inagaki N, Hahnel C, Lindholm D (1993) BDNF and NT-3 induce intracellular Ca^{2+} elevation in hippocampal neurones. NeuroReport 4:1303–1306

Berninger B, Marty S, Zafra F, Berzaghi MP, Thoenen H, Lindholm D (1995) GABAergic stimulation switches from enhancing to repressing BDNF expression in rat hippocampal neurons during maturation in vitro. Development 121:2327–2335

Berzaghi MP, Cooper J, Castren E, Zafra F, Sofroniew G, Thoenen H, Lindholm D (1993) Cholinergic regulation of brain-derived neurotrophic factor (BDNF) and nerve growth factor (NGF)

but not neurotrophin-3 (NT-3) mRNA levels in the developing rat hippocampus. J Neurosci 13:3818–3826
Bhattacharyya A, Watson FL, Bradlee TA, Pomeroy SL, Stiles CD, Segal RA (1997) Trk receptors function as rapid retrograde signal carriers in the adult nervous system. J Neurosci 17:7007–7016
Blochl A, Thoenen H (1995) Characterization of nerve growth factor (NGF) release from hippocampal neurons: evidence for a constitutive and an unconventional sodium-dependent regulated pathway. Eur J Neurosci 7:1220–1228
Blochl A, Thoenen H (1996) Localization of cellular storage compartments and sites of constitutive and activity-dependent release of nerve growth factor (NGF) in primary cultures of hippocampal neurons. MCN 7:173–190
Bolton MM, Pittman AJ, Lo DC (2000) Brain-derived neurotrophic factor differentially regulates excitatory and inhibitory synaptic transmission in hippocampal cultures. J Neurosci 20:3221–3232
Bonhoeffer T (1996) Neurotrophins and activity-dependent development of the neocortex. Curr Opin Neurobiol 6:119–126
Boulanger L, Poo M-M (1999a) Presynaptic depolarization facilitates neurotrophin-induced synaptic potentiation. Nat Neurosci 2:346–351
Boulanger L, Poo M-M (1999b) Gating of BDNF-induced synaptic potentiation by cAMP. Science 284:1982–1984
Bozzi Y, Pizzorusso T, Cremisi F, Rossi FM, Barsacchi G, Maffei L (1995) Monocular deprivation decreases the expression of messenger RNA for brain-derived neurotrophic factor in the rat visual cortex. Neuroscience 69:1133–1144
Cabelli RJ, Hohn A, Shatz CJ (1995) Inhibition of ocular dominance column formation by infusion of NT-4/5 or BDNF. Science 267:1662–1666
Cabelli RJ, Allendoerfer KL, Radeke MJ, Welcher AA, Feinstein SC, Shatz CJ (1996) Changing patterns of expression and subcellular localization of TrkB in the developing visual system. J Neurosci 16:7965–7980
Cabelli RJ, Shelton DL, Segal RA, Shatz CJ (1997) Blockade of endogenous ligands of TrkB inhibits formation of ocular dominance columns. Neuron 19:63–76
Canossa M, Griesbeck O, Berninger B, Campana G, Kolbeck R, Thoenen H (1997) Neurotrophin release by neurotrophins: implications for activity-dependent neuronal plasticity. Proc Natl Acad Sci USA 94:13279–13286
Capsoni S, Tongiorgi E, Cattaneo A, Domenici L (1999) Differential regulation of brain-derived neurotrophic factor messenger mRNA cellular expression in the adult rat visual cortex. Neuroscience 93:1033–1040
Carmignoto G, Canella R, Candeo P, Comelli MC, Maffei L (1993) Effects of nerve growth factor on neuronal plasticity of the kitten visual cortex. J Physiol (Lond) 464:343–360
Carmignoto G, Pizzorusso T, Tia S, Vicini S (1997) Brain-derived neurotrophic factor and nerve growth factor potentiate excitatory synaptic transmission in the rat visual cortex. J Physiol (Lond) 498:153–164
Castellani V, Bolz J (1999) Opposing roles for neurotrophin-3 in targeting and collateral formation of distinct sets of developing cortical neurons. Development 126:3335–3345
Castren E, Zafra F, Thoenen H, Lindholm D (1992) Light regulates expression of brain-derived neurotrophic factor mRNA in rat visual cortex. Proc Natl Acad Sci USA 89:9444–9448
Castren E, Pitkanen M, Sirvio J, Parsadanian A, Lindholm D, Thoenen H, Riekkinen PJ (1993) The induction of LTP increases BDNF and NGF mRNA but decreases NT-3 mRNA in the dentate gyrus. NeuroReport 4:895–898
Causing CG, Gloster A, Aloyz R, Bamji SX, Chang E, Fawcett J, Kuchel G, Miller FD (1997) Synaptic innervation density is regulated by neuron-derived BDNF. Neuron 18:257–267
Chao MV (1992) Growth factor signaling: where is the specificity? Cell 68:995–997
Chen G, Kolbeck R, Barde YA, Bonhoeffer T, Kossel A (1999) Relative contribution of endogenous neurotrophins in hippocampal long-term potentiation. J Neurosci 19:7983–7990

Cohen-Cory S, Fraser SE (1995) Effects of brain-derived neurotrophic factor on optic axon branching and remodeling in vivo. Nature 378:192–196
Collin C, Vicario-Abejon C, Rubio ME, Wenthold RJ, McKay RDG, Segal M (2001) Neurotrophins act at presynaptic terminals to activate synapses among cultured hippocampal neurons. Eur J Neurosci 13:1273–1282
Crino P, Eberwine J (1996) Molecular characterization of the dendritic growth cone: regulated mRNA transport and local protein synthesis. Neuron 17:1173–1187
Crozier RA, Black IB, Plummer MR (1999) Blockade of NR2B-containing NMDA receptors prevents BDNF enhancement of glutamatergic transmission in hippocampal neurons. Learn Mem 6:257–66
Crowley JC, Katz LC (2000) Early development of ocular dominance columns. Science 290: 1321–1324
Davies AM (1994) The role of neurotrophins in the developing nervous system. J Neurobiol 25: 1334–1349
Desai NS, Rutherford LC, Turrigiano GG (1999) BDNF regulated the intrinsic excitability of cortical neurons. Learn Mem 6:284–291
Domenici L, Cellerino A, Maffei L (1993) Monocular deprivation effects in the rat visual cortex and lateral geniculate nucleus are prevented by nerve growth factor (NGF). II. Lateral geniculate nucleus. Proc R Soc Lond (Biol) 251:25–31
Domenici L, Cellerino A, Berardi N, Cattaneo A, Maffei L (1994) Antibodies to nerve growth factor (NGF) prolong the sensitive period for monocular deprivation in the rat. NeuroReport 5: 2041–2044
Dragunow M, Beilharz E, Mason B, Lawlor P, Abraham W, Gluckman P (1993) Brain-derived neurotrophic factor expression after long-term potentiation. Neurosci Lett 160:232–236
Drake CT, Milner TA, Patterson SL (1999) Ultrastructural localization of full-length trkB immunoreactivity in rat hippocampus suggests multiple roles in modulating activity-dependent synaptic plasticity. J Neurosci 19:8009–8026
Du J, Feng L, Yang F, Lu B (2000) Activity- and Ca(2+)-dependent modulation of surface expression of brain-derived neurotrophic factor receptors in hippocampal neurons. J Cell Biol 150: 1423–1434
Dugich-Djordjevic MM, Tocco G, Willoughby DA, Najm I, Pasinetti G, Thompson RF, Baudry M, Lapchak PA, Hefti F (1992) BDNF mRNA expression in the developing rat brain following kainic acid-induced seizure activity. Neuron 8:1127–1138
Ehlers M, Kaplan D, Price D, Koliatsos V (1995) NGF-stimulated retrograde transport of trkA in the mammalian nervous system. J Cell Biol 130:149–156
Eide FF, Lowenstein DH, Reichardt LF (1993) Neurotrophins and their receptors – current concepts and implications for neurologic disease. Exp Neurol 121:200–214
Elmer E, Kokaia M, Kokaia Z, Ferencz I, Lindvall O (1996) Delayed kindling development after rapidly recurring seizures: relation to mossy fiber sprouting and neurotrophin, GAP-43 and dynorphin gene expression. Brain Res 712:19–34
Ernfors P, Ibanez CF, Ebendal T, Olson L, Persson H (1990) Molecular cloning and neurotrophic activities of a protein with structural similarities to nerve growth factor: developmental and topographical expression in the brain. Proc Natl Acad Sci USA 87:5454–5458
Ernfors P, Bengzon J, Kokaia Z, Persson H, Lindvall O (1991) Increased levels of messenger RNAs for neurotrophic factors in the brain during kindling epileptogenesis. Neuron 7:165–176
Fawcett JP, Aloyz R, McLean JH, Pareek S, Miller FD, McPherson PS, Murphy RA (1997) Detection of brain-derived neurotrophic factor in a vesicular fraction of brain synaptosomes. J Biol Chem 272:8837–8840
Figurov A, Pozzo ML, Olafsson P, Wang T, Lu B (1996) Regulation of synaptic responses to high-frequency stimulation and LTP by neurotrophins in the hippocampus. Nature 381:706–709
Fox K, Zahs K (1994) Critical period control in sensory cortex. Curr Opin Neurobiol 4:112–119
Frerking M, Malenka RC, Nicoll RA (1998) Brain-derived neurotrophic factor (BDNF) modulates inhibitory, but not excitatory, transmission in the CA1 region of the hippocampus. J Neurophysiol 80:3383–3386

Frey U, Krug M, Reymann KG, Matthies H (1988) Anisomycin, an inhibitor of protein synthesis, blocks late phases of LTP phenomena in the hippocampal CA region in vitro. Brain Res 452: 57–65

Friedman WJ, Ernfors P, Persson H (1991) Transient and persistent expression of NT-3/HDNF mRNA in the rat brain during postnatal development. J Neurosci 11:1577–1584

Gall CM, Isackson PJ (1989) Limbic seizures increase neuronal production of messenger RNA for nerve growth factor. Science 245:758–761

Galuske RA, Kim DS, Castren E, Thoenen H, Singer W (1996) Brain-derived neurotrophic factor reversed experience-dependent synaptic modifications in kitten visual cortex. Eur J Neurosci 8:1554–1559

Galuske RA, Kim DS, Castren E, Singer W (2000) Differential effects of neurotrophins on ocular dominance plasticity in developing and adult cat visual cortex. Eur J Neurosci 12:3315–3330

Gonzalez M, Ruggiero FP, Chang Q, Shi Y-J, Rich MM, Kraner S, Balice-Gordon RJ (1999) Disruption of TrkB-mediated signaling induces disassembly of postsynaptic receptor clusters at neuromuscular junctions. Neuron 24:567–583

Goodman LJ, Valverde J, Lim F, Geschwind MD, Federoff HJ, Geller AI, Hefti F (1996) Regulated release and polarized localization of brain-derived neurotrophic factor in hippocampal neurons. MCN 7:222–238

Gu Q, Liu Y, Cynader MS (1994) Nerve growth factor-induced ocular dominance plasticity in adult cat visual cortex. Proc Natl Acad Sci USA 91:8408–8412

Gunderson RW, Barrett JN (1979) Neuronal chemotaxis: chick dorsal-root axons turn toward high concentrations of nerve growth factor. Science 206:1079–1080

Hall J, Thomas KL, Everitt BJ (2000) Rapid and selective induction of BDNF expression in the hippocampus during contextual learning. Nat Neurosci 3:533–535

Hata Y, Ohshima M, Ichisaka S, Wakita M, Fukuda M, Tsumoto T (2000) Brain-derived neurotrophic factor expands ocular dominance columns in visual cortex in monocularly deprived and nondeprived kittens but does not in adult cats. J Neurosci 20:RC57

Haubensack W, Narz F, Heumann R, Lessmann V (1998) BDNF-GFP containing secretory granules are localized in the vicinity of synaptic junctions of cultured cortical neurons. J Cell Sci 111:1483–1493

Hebb DO (1949) The organization of behavior: a neuropsychological theory. Wiley, New York

Horch HW, Kruttgen A, Portbury SD, Katz LC (1999) Destabilization of cortical dendrites and spines by BDNF. Neuron 23:353–364

Huang ZJ, Kirkwood A, Pizzorusso T, Porciatti V, Morales B, Bear MF, Maffei L, Tonegawa S (1999) BDNF regulates the maturation of inhibition and the critical period of plasticity in mouse visual cortex. Cell 98:739–755

Hubel DH, Wiesel TN (1970) The period of susceptibility to the physiological effects of unilateral eye closure in kittens. J Physiol (Lond) 206:419–436

Huber KM, Sawtell NB, Bear MF (1998) Brain-derived neurotrophic factor alters the synaptic modification threshold in visual cortex. Neuropharm 37:571–579

Inoue A, Sanes JR (1997) Lamina-specific connectivity in the brain: regulation by N-cadherin, neurotrophins, and glycoconjugates. Science 276:1428–1431

Isackson PJ, Huntsman MM, Murray KD, Gall CM (1991) BDNF mRNA expression is increased in adult rat forebrain after limbic seizures: temporal patterns of induction distinct from NGF. Neuron 6:937–948

Jarvis CR, Xiong ZG, Plant JR, Churchill D, Lu WY, MacVicar BA, MacDonald JF (1997) Neurotrophin modulation of NMDA receptors in cultured murine and isolated rat neurons. J Neurophysiol 78:2363–2371

Jovanovic JN, Benfenati F, Siow YL, Sihra TS, Sanghera JS, Sanghera JS, Pelech SL, Greengard P, Czernik AJ (1996) Neurotrophins stimulate phosphorylation of synapsin I by MAP kinase and regulate synapsin I-actin interactions. Proc Natl Acad Sci USA 93:3679–3683

Kafitz KW, Rose CR, Thoenen H, Konnerth A (1999) Neurotrophin-evoked rapid excitation through TrkB receptors. Nature 401:918–921

Kang H, Schuman EM (1995) Long-lasting neurotrophin-induced enhancement of synaptic transmission in the adult hippocampus. Science 267:1658–1662
Kang H, Schuman EM (1996) A requirement for local protein synthesis in neurotrophin-induced hippocampal synaptic plasticity. Science 273:1402–1406
Kang H, Welcher AA, Shelton D, Schuman EM (1997) Neurotrophins and time: different roles for TrkB signaling in hippocampal long-term potentiation. Neuron 19:653–664
Kaplan DR, Miller FD (2000) Neurotrophin signal transduction in the nervous system. Curr Opin Neurobiol 10:381–391
Katz LC, Shatz CJ (1996) Synaptic activity and the construction of cortical circuits. Science 274: 1133–1138
Kim HG, Wang T, Olafsson P, Lu B (1994) Neurotrophin-3 potentiates neuronal activity and inhibits gamma-aminobutyratergic synaptic transmission in cortical neurons. Proc Natl Acad Sci USA 91:12341–12345
Klein R, Conway D, Parada LF, Barbacid M (1990a) The trkB tyrosine kinase gene codes for a second neurogenic receptor that lacks the catalytic kinase domain. Cell 61:647–656
Klein R, Martin-Zanca D, Barbacid M, Parada LF (1990b) Expression of the tyrosine kinase receptor gene trkB is confined to the murine embryonic and adult nervous system. Development 109:845–850
Knusel B, Rabin SJ, Hefti F, Kaplan DR (1994) Regulated neurotrophin receptor responsiveness during neuronal migration and early differentiation. J Neurosci 14:1542–1554
Kohara K, Litamura A, Morishima M, Tsumoto T (2001) Activity-dependent transfer of brain-derived neurotrophic factor to postsynaptic neurons. Science 291:2419–2423
Kokaia Z, Bengzon J, Metsis M, Kokaia M, Persson H, Lindvall O (1993) Coexpression of neurotrophins and their receptors in neurons of the central nervous system. Proc Natl Acad Sci USA 90:6711–6715
Korte M, Carroll P, Wolf E, Brem G, Thoenen H, Bonhoeffer T (1995) Hippocampal long-term potentiation is impaired in mice lacking brain-derived neurotrophic factor. Proc Natl Acad Sci USA 92:8856–8860
Korte M, Griesbeck O, Gravel C, Carroll P, Staiger V, Thoenen H, Bonhoeffer T (1996) Virus-mediated gene transfer into hippocampal CA1 region restores long-term potentiation in brain-derived neurotrophic factor mutant mice. Proc Natl Acad Sci USA 93:12547–12552
Kryl D, Yacoubian T, Haapasalo A, Castren E, Lo D, Barker P (1999) Subcellular localization of full-length and truncated Trk receptor isoforms in polarized neurons and epithelial cells. J Neurosci 19:5823–5833
Lauterborn JC, Lynch G, Vanderklish P, Arai A, Gall CM (2000) Positive modulation of AMPA receptors increases neurotrophin expression by hippocampal and cortical neurons. J Neurosci 20:8–21
Lessmann V, Gottmann K, Heumann R (1994) BDNF and NT-4/5 enhance glutamatergic synaptic transmission in cultured hippocampal neurones. Neuroreport 6:21–25
Lein ES, Shatz CJ (2000) Rapid regulation of brain-derived neurotrophic factor mRNA within eye-specific circuits during ocular dominance column formation. J Neurosci 20:1470–1483
Lein ES, Finney EM, McQuillen PS, Shatz CJ (1999) Subplate neuron ablation alters neurotrophin expression and ocular dominance column formation. Proc Natl Acad Sci USA 96:13491–13495
Levi-Montalcini R (1987) The nerve growth factor 35 years later. Science 237:1154–1162
Levi-Montalcini R, Meyer H, Hamburger V (1954) In vitro experiments on the effects of mouse sarcomas 180 and 37 on the spinal and sympathetic ganglia of the chick embryo. Cancer Res 14:49–57
Levine ES, Dreyfus CF, Black IB, Plummer MR (1995) Brain-derived neurotrophic factor rapidly enhances synaptic transmission in hippocampal neurons via postsynaptic tyrosine kinase receptors. Proc Natl Acad Sci USA 92:8074–8077
Li YX, Xu Y, Ju D, Lester HA, Davidson N, Schuman EM (1998a) Expression of a dominant negative TrkB receptor, T1, reveals a requirement for presynaptic signaling in BDNF-induced synaptic potentiation in cultured hippocampal neurons. Proc Natl Acad Sci USA 95:10884–10889

Li YX, Zhang Y, Lester HA, Schuman EM, Davidson N (1998b) Enhancement of neurotransmitter release induced by brain-derived neurotrophic factor in cultured hippocampal neurons. J Neurosci 18:10231–10240

Lindholm D, Castren E, Berzaghi MP, Blochl A, Thoenen H (1994) Activity-dependent and hormonal regulation of neurotrophin mRNA levels in the brain – implications for neuronal plasticity. J Neurobiol 25:1362–1372

Lindsay RM, Wiegand SJ, Altar CA, DiStefano PS (1994) Neurotrophic factors: from molecule to man. Trends Neurosci 17:182–190

Lo DC (1995) Neurotrophic factors and synaptic plasticity. Neuron 15:979–981

Lodovichi C, Berardi N, Pizzorusso T, Maffei L (2000) Effects of neurotrophins on cortical plasticity: same or different? J Neurosci 20:2155–2165

Loeb JA, Fischbach GD (1997) Neurotrophic factors increase neuregulin expression in embryonic ventral spinal cord neurons. J Neurosci 17:1416–1424

Lohof AM, Ip NY, Poo M-M (1993) Potentiation of developing neuromuscular synapses by the neurotrophins NT-3 and BDNF. Nature 363:350–352

Lu B, Yokoyama M, Dreyfus CF, Black IB (1991) Depolarizing stimuli regulate nerve growth factor gene expression in cultured hippocampal neurons. Proc Natl Acad Sci USA 88:6289–6292

MacKay DJ, Hall A (1998) Rho GTPases. J Biol Chem 273:20685–20688

Maffei L, Berardi N, Domenici L, Parisi V, Pizzorusso T (1992) Nerve growth factor (NGF) prevents the shift in ocular dominance distribution of visual cortical neurons in monocularly deprived rats. J Neurosci 12:4651–4662

Maisonpierre PC, Belluscio L, Squinto S, Ip NY, Furth ME, Lindsay RM, Yancopoulos GD (1990) Neurotrophin-3: a neurotrophic factor related to NGF and BDNF. Science 247:1446–145

Martinez A, Alcantara S, Borrell V, Rio JAD, Blasi J, Otal R, Campos N, Boronat A, Barbacid M, Silos-Santiago I, Soriano E (1998) TrkB and TrkC signaling are required for maturation and synaptogenesis of hippocampal connections. J Neurosci 18:7336–7350

Marty S, Berninger B, Carroll P, Thoenen H (1996a) GABAergic stimulation regulates the phenotype of hippocampal interneurons through the regulation of brain-derived neurotrophic factor. Neuron 16:565–570

Marty S, Carroll P, Cellerino A, Castren E, Staiger V, Thoenen H, Lindholm D (1996b) Brain-derived neurotrophic factor promotes the differentiation of various hippocampal nonpyramidal neurons, including Cajal-Retzius cells, in organotypic slice cultures. J Neurosci 16:675–687

McAllister AK (2000) Cellular and molecular mechanisms of dendrite growth. Cerebral Cortex 10:963–973

McAllister AK (2001) Neurotrophins and neuronal differentiation in the central nervous system. Cell Mol Life Sci 58:1054–1060

McAllister AK, Lo DC, Katz LC (1995) Neurotrophins regulate dendritic growth in developing visual cortex. Neuron 15:791–803

McAllister AK, Katz LC, Lo DC (1996) Neurotrophin regulation of cortical dendritic growth requires activity. Neuron 17:1057–1064

McAllister AK, Katz LC, Lo DC (1997) Opposing roles for endogenous BDNF and NT-3 in regulating cortical dendritic growth. Neuron 18:767–778

McAllister AK, Katz LC, Lo DC (1999) Neurotrophins and synaptic plasticity. Annu Rev Neurosci 22:295–318

Meyer-Franke A, Wilkinson GA, Kruttgen A, Hu M, Munro E, Hanson MG Jr, Reichardt LF, Barres BA (1998) Depolarization and cAMP elevation rapidly recruit TrkB to the plasma membrane of CNS neurons. Neuron 21:681–693

Middlemas DS, Lindberg RA, Hunter T (1991) trkB, a neural receptor protein-tyrosine kinase: evidence for a full-length and two truncated receptors. Mol Cell Biol 11:143–153

Ming GI, Lohof AM, Zheng JQ (1997) Acute morphogenic and chemotropic effects of neurotrophins on cultured embryonic Xenopus spinal neurons. J Neurosci 17:7860–7871

Mudo G, Jiang XH, Timmusk T, Bindoni M, Belluardo N (1996) Change in neurotrophins and their receptor mRNAs in the rat forebrain after status epilepticus induced by pilocarpine. Epilepsia 37:198–207

Nguyen PV, Abel T, Kandel ER (1994) Requirement of a critical period of transcription for induction of a late phase of LTP. Science 265:1104–1107
Oppenheim RW (1991) Cell death during development of the nervous system. Annu Rev Neurosci 14:453–501
Patterson SL, Grover LM, Schwartzkroin PA, Bothwell M (1992) Neurotrophin expression in rat hippocampal slices: a stimulus paradigm inducing LTP in CA1 evokes increases in BDNF and NT-3 mRNAs. Neuron 9:1081–1088
Patterson SL, Abel T, Deuel TA, Martin KC, Rose JC, Kandel ER (1996) Recombinant BDNF rescues deficits in basal synaptic transmission and hippocampal LTP in BDNF knockout mice. Neuron 16:1137–1145
Paves H, Saarma M (1997) Neurotrophins as in vitro growth cone guidance molecules for embryonic sensory neurons. Cell Tissue Res 290:285–297
Poo M-M (2001) Neurotrophins as synaptic modulators. Nat Rev Neurosci 2:24–32
Pozzo-Miller LD, Gottschalk W, Zhang L, McDermott K, Du J, Gopalakrishnan R, Oho C, Sheng ZH, Lu B (1999) Impairment in high-frequency transmission, synaptic vesicle docking, and synaptic protein distribution in the hippocampus of BDNF knockout mice. J Neurosci 19: 4972–4983
Purves D (1988) Body and brain. Harvard University Press, Cambridge, MA
Riddle DR, Lo DC, Katz LC (1995) NT-4-mediated rescue of lateral geniculate neurons from effects of monocular deprivation. Nature 378:189–191
Righi M, Tongiorgi E, Cattaneo A (2000) Brain-derived neurotrophic factor (BDNF) induces dendritic targeting of BDNF and tyrosine kinase B mRNAs in hippocampal neurons through a phosphatidylinositol-3-kinase-dependent pathway. J Neurosci 20:3165–3174
Ringstedt T, Lagercrantz H, Persson H (1993) Expression of members of the trk family in the developing postnatal rat brain. Brain Res Dev Brain Res 72:119–131
Rocamora N, Welker E, Pascual M, Soriano E (1996) Upregulation of BDNF mRNA expression in the barrel cortex of adult mice after sensory stimulation. J Neurosci 16:4411–4419
Rossi FM, Bozzi Y, Pizzorusso T, Maffei L (1999) Monocular deprivation decreases brain-derived neurotrophic factor immunoreactivity in the rat visual cortex. Neuroscience 90:363–368
Ruit KG, Snider WD (1991) Administration or deprivation of nerve growth factor during development permanently alters neuronal geometry. J Comp Neurol 314:106–131
Ruit KG, Osborne PA, Schmidt RE, Johnson EM Jr, Snider WD (1990) Nerve growth factor regulates sympathetic ganglion cell morphology and survival in the adult mouse. J Neurosci 10:2412–2419
Rutherford LO, DeWan A, Lauer HM, Turrigiano GG (1997) Brain-derived neurotrophic factor mediates the activity-dependent regulation of inhibition in neocortical cultures. J Neurosci 17:4527–4535
Sala R, Viegi A, Rossi FM, Pizzorusso T, Bonanno G, Raiteri M, Maffei L (1998) Nerve growth factor and brain-derived neurotrophic factor increase neurotransmitter release in rat visual cortex. Eur J Neurosci 10:2185–2191
Schoups AA, Elliott RC, Friedman WJ, Black IB (1995) NGF and BDNF are differentially modulated by visual experience in the developing geniculocortical pathway. Brain Res 86:326–334
Seil FJ, Drake-Baumann R (2000) TrkB receptor ligands promote activity-dependent inhibitory synaptogenesis. J Neurosci 20:5367–5373
Shatz CJ (1983) The prenatal development of the cat's retinogeniculate pathway. J Neurosci 3: 482–499
Shatz CJ (1990) Impulse activity and the patterning of connections during CNS development. Neuron 5:745–756
Sherwood NT, Lo DC (1999) Long-term enhancement of central synaptic transmission by chronic brain-derived neurotrophic factor treatment. J Neurosci 19:7025–7036
Smith MA, Zhang LX, Lyons WE, Mamounas L (1997) Anterograde transport of endogenous brain-derived neurotrophic factor in hippocampal mossy fibers. Neuroreport 8:1829–1834
Snider WD (1988) Nerve growth factor enhances dendritic arborizations of sympathetic ganglion cells in developing mammals. J Neurosci 8:2628–2634

Snider WD (1994) Functions of the neurotrophins during nervous system development: what the knockouts are teaching us. Cell 77:627–638
Snider WD, Lichtman JW (1996) Are neurotrophins synaptotrophins? MCN 7:433–442
Song HJ, Ming GL, Poo M-M (1997) cAMP-induced switching in turning of nerve growth cones. Nature 388:275–279
Stoop R, Poo M-M (1996) Synaptic modulation by neurotrophic factors: differential and synergistic effects of brain-derived neurotrophic factor and ciliary neurotrophic factor. J Neurosci 16:3256–3264
Suen PC, Wu K, Levine ES, Mount HT, Xu JL, Lin SY, Black IB (1997) Brain-derived neurotrophic factor rapidly enhances phosphorylation of the postsynaptic N-methyl-D-aspartate receptor subunit 1. Proc Natl Acad Sci USA 94:8191–8195
Takei N, Sasaoka K, Inoue K, Takahashi M, Endo Y, Hatanaka H (1997) Brain-derived neurotrophic factor increases the stimulation-evoked release of glutamate and the levels of exocytosis-associated proteins in cultured cortical neurons from embryonic rats. J Neurochem 68:370–375
Tanaka T, Saito H, Matsuki N (1997) Inhibition of $GABA_A$ synaptic responses by brain-derived neurotrophic factor (BDNF) in rat hippocampus. J Neurosci 17:2959–2966
Thoenen H (1995) Neurotrophins and neuronal plasticity. Science 270:593–598
Timmusk T, Belluardo N, Metsis M, Persson H (1993) Widespread and developmentally regulated expression of neurotrophin-4 mRNA in rat brain and peripheral tissues. Eur J Neurosci 5: 605–613
Tongiorgi E, Righi M, Cattaneo A (1997) Activity-dependent dendritic targeting of BDNF and TrkB mRNAs in hippocampal neurons. J Neurosci 17:9492–9505
Tsoulfas P, Soppet D, Escandon E, Tessarollo L, Mendoza-Ramirez JL, Rosenthal A, Nikolics K, Parada LF (1993) The rat trkC locus encodes multiple neurogenic receptors that exhibit differential response to neurotrophin-3 in PC12 cells. Neuron 10:975–990
Valenzuela DM, Maisonpierre PC, Glass DJ, Rojas E, Nunez L, Kong Y, Gies DR, Stitt TN, Ip NY, Yancopoulos GD (1993) Alternative forms of rat TrkC with different functional capabilities. Neuron 10:963–974
Vicario-Abejon C, Collin C, McKay RDG, Segal M (1998) Neurotrophins induce formation of functional excitatory and inhibitory synapses between cultured hippocampal neurons. J Neurosci 18:7256–7271
von Bartheld CS, Butowt R (2000) Expression of neurotrophin-3 (NT-3) and anterograde axonal transport of endogenous NT-3 by retinal ganglion cells in chick embryos. J Neurosci 20: 736–748
von Bartheld CS, Byers MR, Williams R, Bothwell M (1996) Anterograde transport of neurotrophins and axodendritic transfer in the developing visual system. Nature 379:830–833
Wang T, Xie KW, Lu B (1995) Neurotrophins promote maturation of developing neuromuscular synapses. J Neurosci 15:4796–4805
Wang Q, Zheng JQ (1998) cAMP-mediated regulation of neurotrophin-induced collapse of nerve growth cones. J Neurosci 18:4973–4984
Wang X, Poo M-M (1997) Potentiation of developing synapses by postsynaptic release of neurotrophin-4. Neuron 19:825–835
Watson FL, Heerssen HM, Moheban DB, Lin MZ, Sauvageot CM, Bhattacharyya A, Pomeroy SL, Segal RA (1999) Rapid nuclear responses to target-derived neurotrophins require retrograde transport of ligand-receptor complex. J Neurosci 19:7889–7900
Wells DG, McKechnie BA, Kelkar S, Fallon JR (1999) Neurotrophins regulate agrin-induced postsynaptic differentiation. Proc Natl Acad Sci USA 96:1112–1117
Widmer HR, Hefti F (1994) Stimulation of GABAergic neuron differentiation by NT4/5 in cultures of rat cerebral cortex. Dev Brain Res 80:279–284
Wu K, Xu JL, Suen PC, Levine E, Huang YY, Mount HT, Lin SY, Black IB (1996) Functional trkB neurotrophin receptors are intrinsic components of the adult brain postsynaptic density. Brain Res Mol Brain Res 43:286–290
Xu B, Gottschalk W, Chow A, Wilson RI, Schnell E, Zang K, Wang D, Nicoll RA, Lu B, Reichardt LF (2000a) The role of brain-derived neurotrophic factor receptors in the mature hippocam-

pus: modulation of long-term potentiation through a presynaptic mechanism involving TrkB. J Neurosci 20:6888–6897

Xu B, Zang K, Ruff NL, Zhang YA, McConnell SK, Stryker MP, Reichardt LF (2000b) Cortical degeneration in the absence of neurotrophin signaling: dendritic retraction and neuronal loss after removal of the receptor TrkB. Neuron 26:233–245

Yacoubian TA, Lo DC (2000) Truncated and full-length TrkB receptors regulate distinct modes of dendritic growth. Nat Neurosci 3:42–349

Yamashita T, Tucker KL, Barde YA (1999) Neurotrophin binding to the p75 receptor modulates Rho activity and axonal outgrowth. Neuron 24:585–593

Yan Q, Rosenfeld RD, Matheson CR, Hawkins N, Lopez OT, Bennett L, Welcher AA (1997) Expression of brain-derived neurotrophic factor (BDNF) protein in the adult rat central nervous system. Neuroscience 78:431–448

Zafra F, Hengerer B, Leibrock J, Thoenen H, Lindholm D (1990) Activity-dependent regulation of BDNF and NGF mRNA in the rat hippocampus is mediated by non-NMDA glutamate receptors. EMBO J 9:3545–3550

Zafra F, Castren E, Thoenen H, Lindholm D (1991) Interplay between glutamate and gamma-aminobutyric acid transmitter systems in the physiological regulation of brain-derived neurotrophic factor and nerve growth factor synthesis in hippocampal neurons. Proc Natl Acad Sci USA 88:10037–10041

Zafra F, Lindholm D, Castren E, Hartikka J, Thoenen H (1992) Regulation of brain-derived neurotrophic factor and nerve growth factor mRNA in primary cultures of hippocampal neurons and astrocytes. J Neurosci 12:4793–4799

Role of Immediate Early Gene Expression in Cortical Morphogenesis and Plasticity

KATRIN I. ANDREASSON[1] and WALTER E. KAUFMANN[2]

Summary. During the development of the central nervous system, there is a fundamental requirement for synaptic activity in transforming immature neuronal connections into organized functional circuits (Katz 1996). The molecular mechanisms underlying activity-dependent adaptive changes in neurons are believed to involve regulated cascades of gene expression. Immediate early genes (IEGs) comprise the initial cascade of gene expression responsible for initiating the process of stimulus-induced adaptive change, and were identified initially as transcription factors that were regulated in brain by excitatory synaptic activity. More recently, a class of neuronal immediate early genes has been identified that encodes growth factors, signaling molecules, extracellular matrix and adhesion proteins, and cytoskeletal proteins that are rapidly and transiently expressed in response to glutamatergic neurotransmission. This review focuses on the neuronal immediate early gene (nIEG) response, in particular, the class of "effector" immediate early gene proteins that may directly modify neuronal and synaptic function.

1 Neural Activity Plays a Critical Role in the Development of the Cerebral Cortex

In most mammals, several brain regions are not fully developed at birth, most notably, the cerebral cortex (Dobbing and Sands 1979; reviewed by Kaufmann 2001). Although in most species cortical neurons are already generated at birth, these cells are at relatively early stages of cytodifferentiation (Petit et al. 1988). Neocortical synaptic formation and remodeling are essentially postnatal phenomena in rodents and humans (McConnell 1988; Mrzljak et al. 1990). Other regressive phenomena, such as programmed cell death in central nervous system (CNS) regions involved in higher integrative functions, also occur mainly postnatally (Raff et al. 1993). Consequently, the "adult" configuration of the cerebral cortex is acquired during early postnatal life by selective preservation and elimination of neurons and their connections, guided by extracellular influences.

Neuronal impulse activity appears to be essential for transforming immature neuronal connections into organized functional circuits (Constantine-

[1] Departments of Neurology and Neuroscience, Johns Hopkins University School of Medicine, 600 N. Wolfe Street, Meyer 5-119B, Baltimore, Maryland 21205, USA

[2] Departments of Neurology, Neuroscience, Pathology, Pediatrics, Psychiatry and Behavioral Studies, and Radiological Science, Johns Hopkins University School of Medicine and the Kennedy Krieger Institute, 600 N. Wolfe Street, Meyer 5-119B, Baltimore, Maryland 21205, USA

Results and Problems in Cell Differentiation, Vol. 39
C.F. Hohmann (Ed.): Cortical Development

Paton et al. 1990; Shatz 1990; Goodman and Shatz 1993; Katz and Shatz 1996). The developmental role of synaptic activity has been most extensively studied in the visual system. Many current concepts that relate sensory experience to the correct formation of neural circuits are based on the pioneering studies of Hubel and Wiesel (Wiesel 1970; Hubel et al. 1977), who found that deprivation of visual input in the neonatal cat dramatically disrupted the normal formation of ocular dominance columns in the visual cortex. Recent studies focusing on the development of the rodent somatosensory system have also demonstrated an association between afferent synaptic activity and proper formation of the barrel field cortex (Schlaggar et al. 1993).

Interestingly, neural activity not only modulates the connectivity of sensory pathways, but also events that precede postnatal sensory experience (Katz and Shatz 1996). In the visual system, spontaneous neuronal activity generated before the onset of sensory experience is necessary for proper synaptic wiring. Subsequent postnatal sensory stimulation is postulated to refine and amplify existing synaptic contacts. For example, spontaneous bursts of retinal ganglion cell activity occur long before the onset of vision and promote proper development of eye-specific layers in the lateral geniculate nucleus of the thalamus (Shatz and Stryker 1988; Maffei and Galli-Resta 1990; Meister et al. 1991; Wong et al. 1993). Moreover, spontaneously generated retinal action potentials drive geniculate neurons to fire periodic bursts of action potentials, and this neural activity may influence synaptic modifications within the developing visual cortex (Mooney et al. 1996).

Further evidence supporting the existence of pre-experiential activity-dependent processes in the visual system includes the demonstration that formation of ocular dominance columns occurs prenatally in the monkey (Rakic 1976) and is complete at birth (Horton and Howking 1996). Neural activity appears to also play a modulatory role during even earlier developmental stages. Correlated spontaneous activity has been observed in embryonic chick lumbar motor neurons (O'Donovan et al. 1992) and among clusters of developing cortical neurons connected to one another by gap junctions (Yuste et al. 1992, 1995). These interactions may occur as early as the period of neuronal proliferation and migration. For instance, neuroblasts coupled by gap junctions in the rat ventricular zone can generate large responses to inhibitory neurotransmitters (LoTurco et al. 1991). Other examples of postmigrational prenatal neural activity are the responses of *N*-methyl-D-aspartate (NMDA) receptors on differentiating neurons of embryonic cortical plate to endogenous neurotransmitters (Blanton et al. 1990), and the transient synaptic interactions in the subplate layer that determine the establishment of proper thalamocortical connections in the developing cortex (Ghosh and Shatz 1992).

The identification of the physiological and molecular underpinnings that underlie activity-dependent neuronal plasticity is essential for attaining a basic understanding of both cortical development and neurological developmental disorders. For example, cerebral palsy resulting from perinatal hypoxic-ischemic injury involves a dramatic reorganization of the cerebral cortex that allows the child to use areas of the intact brain for other functions not normally subserved by those areas (Ono et al. 1991). Conversely, nonadaptive plastic responses may be responsible for some of the motor impairments found in cerebral palsy (Farmer et al. 1991). Aspects of this adaptive process are likely to involve activity-dependent plasticity and might be amenable to therapies that could modify the process. The range of disorders in which abnormal cerebral cortex formation may result from aberrant activity-dependent developmental processes includes, among others, genetic and environmental conditions associated with mental retardation (Huttenlocher 1991; Kaufmann and Moser 2000) and schizophrenia (Gruzellier and Flor-Henry 1979). The latter has been recently linked to abnormalities in the organization of the subplate layer (Akbarian et al. 1993; Anderson et al. 1996), which may lead to aberrant synaptic interactions during cell migration and neuronal differentiation.

2
Learning and Development Share Mechanisms of Neural Plasticity

The ability of neurons to adapt to a changing environment is defined as neuronal plasticity. In many experimental models, plasticity is known to be dependent on synaptic activity and subsequent neurotransmitter activation of postsynaptic receptors. In recent years, it has become evident that many of the mechanisms involved in plastic phenomena in the adult brain also participate in the establishment and refinement of brain connections. For instance, developmental plasticity in the visual cortex has been shown to share a number of characteristics with long-term potentiation (LTP), an extensively studied paradigm of activity-dependent plasticity in the mature hippocampus (Artola and Singer 1987; Fox and Daw 1993). In slice preparations of the CA1 hippocampus, high-frequency trains of electrical stimuli, which result in the coordinated depolarization of both pre- and postsynaptic neurons, strengthen synapses for many hours. Induction of LTP has been shown to be dependent on activation of NMDA-type glutamate receptors, which are notable for having both a voltage (Mg^{2+} comes off with depolarization) and a ligand (glutamate) activation dependency. Depolarization of the pre- and postsynaptic neurons has been shown to be necessary in visual cortical synaptic plasticity (Reiter et al. 1986; Reiter and Stryker 1988), implicating the activation of NMDA receptors in synaptic plasticity of developing (Tsumoto et al. 1987; Fox et al. 1989) and adult cats (Kleinschmidt et al. 1987; Hagihara et al. 1988; Miller et al. 1989; Bear et al. 1990). In the developing neocortex, LTP may contribute to the segregation of ocular dominance columns (Artola and Singer 1987; Tsumoto 1992) and has been demonstrated in thalamocortical connections of the rodent somatosensory system during early postnatal life (Crair and Malenka 1995) and in the local connections of the visual cortex between layers IV and II/III in young and old animals (Kirkwood and Bear 1994; Kirkwood et al. 1995). The physiological parallels between LTP and developmental cortical plasticity support the hypothesis that molecular mechanisms involved in LTP and developmental plasticity may overlap.

3
Molecular Events Underlying Cortical Plasticity: the Immediate Early Gene Response

While the initial events induced by transmitter stimulation are well characterized, mechanisms underlying long-term structural and physiological changes remain conjectural but are believed to involve the activation of specific genes. These molecular pathways linking neuronal depolarization and synapse strengthening, formation, or elimination appear to involve prominently among others the NMDA-type of glutamate receptor. NMDA activation results in rapid ionic fluxes and subsequent signaling cascades that may initiate long-term changes in neuronal phenotype (Bliss and Collingridge 1993). Thus calcium, entering through NMDA receptor channels or voltage-gated ion channels, activates calcium-dependent kinases which, in turn, control the phosphorylation state and functional activity of a specific set of proteins that include transcription factors such as CREB (cyclic AMP response element binding protein) and SRF (serum response factor; Bading et al. 1993; Ghosh et al. 1994b). These proteins, which constitute the basal transcription apparatus of neurons, stimulate the expression of an inducible set of genes: the immediate early genes.

The term “immediate early gene” (IEG) refers to the initial cascade of gene expression following an extracellular signal (Fig. 1). IEGs were initially described as part of the proliferative response from fibroblasts to growth factors. These activated genes were found to encode transcription factors whose mRNA levels increased without the need for new protein synthesis (Cochran et al. 1983; Linzer and Nathans 1983). IEG transcription leads to either more transcription factor activation or to the expression of proteins directly involved in altering the phenotype of the stimulated cell (delayed or late genes). These secondary events require the synthesis of new proteins. The obvious parallels between fibroblast stimulation and neuronal responses to synaptic activity led to the examination of IEGs in neural systems. Following initial in vitro studies, using PC12 cells and other simple paradigms, induction of IEGs in brain (Sheng and Greenberg 1990) was demonstrated in standard models of plasticity. Transcription factor expression was rapidly induced in hippocampus and cortex by NMDA receptor-dependent synaptic activity (e.g., LTP), both during development and in the adult brain (Morgan et al. 1987; Curran and Franza 1988; Saffen et al. 1988; Cole et al. 1989; Sheng and Greenberg 1990). Moreover, the developmental profile of several of these IEGs paralleled both glutamate receptor ontogeny and the late phase of dendritic/synaptic formation (Kaufmann et al. 1994). The genes that are rapidly

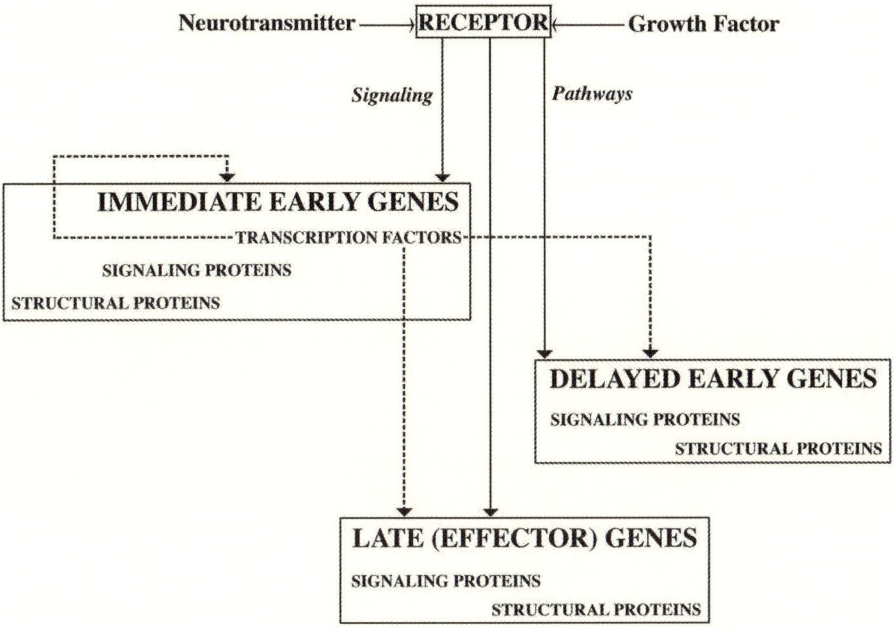

Fig. 1. Model of immediate, delayed and late gene responses in neurons in response to excitatory synaptic activity. (Reprinted courtesy of MRDD Research Reviews, Kaufmann and Worley 1999)

regulated by NMDA-dependent synaptic activity have been termed neuronal immediate early genes (nIEGs). An initial group of nIEGs was identified as encoding a group of transcription factors, including members of the AP-1 transcriptional complex and zinc finger proteins (Kaufmann and Worley 1999). Further characterization of the nIEG response during developmental plasticity and LTP demonstrated a second category of nIEGs that encoded proteins that were not transcription factors, but rather had the potential of directly modifying synaptic connections, and these have been termed "effector" nIEGs. These effector nIEGs encode a wide variety of structural and signaling proteins, including actin-binding and anchoring proteins and enzymes involved in signal transduction (Kaufmann and Worley 1999), which will be discussed in detail in the following sections.

Most early discovered nIEGs were found to regulate transcription, supporting the concept that cellular responses to synaptic activity involve regulated cascades of gene activation (Lau and Nathans 1991). Importantly, these transcription factor nIEGs are regulated by physiological activity (Hunt et al. 1987; Sharp et al. 1991; Worley et al. 1991) and are therefore likely to be part of the natural cellular response to activity. Transcription factor nIEGs are prominently expressed in neurons of the normal adult and developing cortex. Basal expression appears to be driven by normal excitatory synaptic activity since interruption of afferent activity, for example, by infusion of the sodium channel blocker tetrodotoxin into one eye, results in a rapid decrease in cortical basal expression (Worley et al. 1990a; Chaudhuri and Cynader 1993). Basal expression of many of the activity-regulated transcription factors is also dynamically regulated during postnatal development (Fig. 2). Highest levels of transcription factor nIEG expression typically occur during late postnatal development (Worley et al. 1990b; Kaufmann et al. 1994), and coincide with the period of maximal sensitivity of the developing cortex to environmental stimuli (Dobbing and Smart 1974; Greenough 1984). During this "critical period" of postnatal development, brief visual experience induces the expression of transcription factor nIEGs in the visual cortex (Worley et al. 1990b; Rosen et al. 1992) indicating that basal expression during postnatal development is regulated by natural synaptic activity. Figure 2 illustrates the postnatal profile of several transcription factor nIEGs.

As mentioned above, one of the most intriguing concepts in neural plasticity is that the nIEG response includes genes that can directly modify neuronal structure and function. Long-term adaptive changes in neurons result from

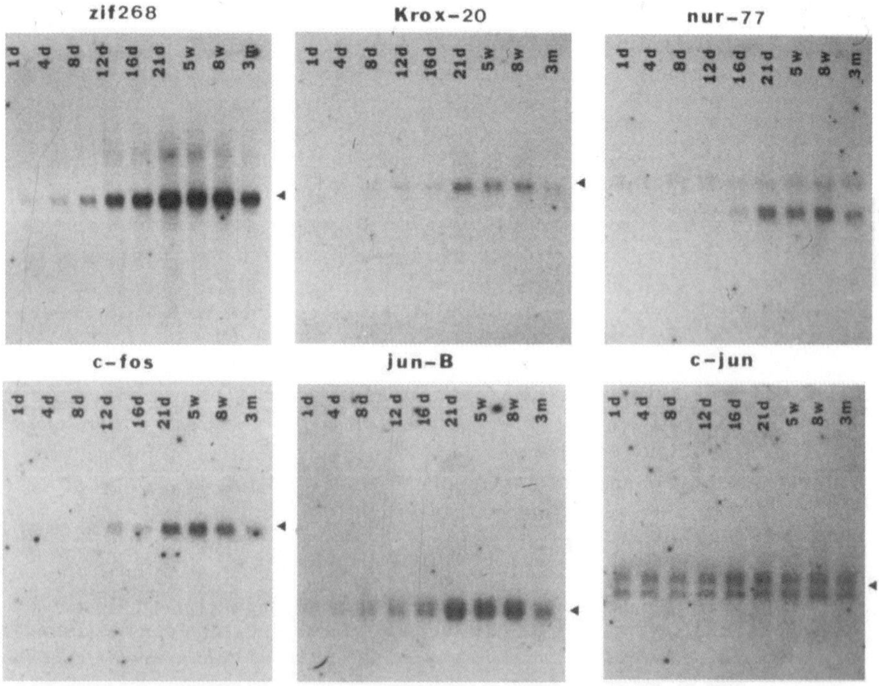

Fig. 2. Northern analysis demonstrating developmental profiles of several nIEG transcription factors during early postnatal development from postnatal day 1 (d1) through 3 months (3 m). (Reprinted courtesy of Int. J. Dev. Neuroscience, Kaufmann et al. 1994)

integration of downstream effects of transcription factor nIEGs and effector nIEGs. Effector nIEGs, like their transcription factor nIEG counterparts, are regulated by NMDA-dependent activity in brain and, at the present time, their function in mediating long-term phenotypic changes is the focus of intense investigation (Kaufmann and Worley 1999). Immunolocalization of several of these effector nIEG proteins to dendrites (Lyford et al. 1995; Tsui et al. 1996) and, specifically, to excitatory synapses (Kaufmann et al. 1996; Brakeman et al. 1997; O'Brien et al. 1999; Yamagata et al. 1999) supports the hypothesis that these nIEGs function in activity-induced changes in synaptic connectivity. More importantly, a number of these effector nIEGs are expressed during early stages of cortical development in specific temporal and spatial distributions. Therefore, effector nIEGs may play an important role in prenatal and postnatal phenomena leading to the establishment of synaptic connections. In the following sections, the term nIEGs will only be used to refer to this novel category of genes, also designated effector nIEGs, which are the focus of this review. We will examine recent studies of representative members of this novel class of nIEGs: the inducible form of prostaglandin synthase (Yamagata et al. 1993), a novel member of the Ras family of GTP-binding regulatory proteins (Yamagata et al. 1994), tissue plasminogen activator (Quian et al. 1993), a member of the transforming-growth factor β family (Andreasson and Worley 1995), a dendrite-enriched cytoskeletal protein enriched (Lyford et al. 1995), a novel member of the pentraxin family of secreted lectins (Tsui et al. 1996), a novel anchoring protein (Brakeman et al. 1997) and a novel member of the cadherin superfamily of adhesion molecules (Yamagata et al. 1999).

4
Effector Neuronal Immediate Early Genes

4.1
Growth Factors: Activin and BDNF

Activin. Activin A is a member of the transforming growth factor β (TGFβ) family of secreted peptides and, like other TGFβ members, has potent effects on cellular growth and differentiation during development. It functions as a dimer composed of either β-A and/or β-B subunits (Mather et al. 1992, 1997) and signals through transmembrane serine-threonine kinase receptors. In the adult brain, expression of activin A is rapidly and transiently regulated by glutamatergic synaptic activity and, in particular, NMDA-receptor mediated synaptic activity (Andreasson and Worley 1995).

Activin A is widely expressed in developing organs, and has a unique pattern of expression in the developing brain (Andreasson and Worley 1995). In the prenatal cortex, during the most active period of histogenesis (e.g., embryonic day 17 in the rat), activin expression is enriched in discrete laminae of cells that are located at the boundary between the hypercellular upper cortical plate and subjacent maturing deep layers. These upper laminae contain immature postmitotic neuroblasts, with a simple bipolar phenotype and no well-defined dendritic arbor, which are beginning the process of cytodifferentiation that is at a more advanced stage in the deeper layers of the cortex. This pattern of neuronal differentiation is the result of the unique way in which the cortex develops. The first group of postmitotic/postmigrational neuroblasts constitutes the preplate layer, the earliest cortical stratum. This lamina then splits into two layers (layer I and the subplate layer; Bayer and Altman 1991). The superficial component becomes the molecular layer or lamina I, a stratum populated by nonpyramidal neurons that is greatly influenced by monoaminergic influences (Molliver and Krist 1975). The subplate layer in turn becomes a major organizer of subsequent cortical development, with neuroblasts destined to future layers II–VI migrating through the subplate in an "inside-outside" fashion. That is, the deeper layers of the cortex (VI, V) develop earlier than the more superficial strata. Activin A expression is first detected in the preplate layer and later is found, as mentioned above, at the boundary between the developing cortical plate (supragranular layers) and the subjacent lamina that will become layers VI and V (Bayer and Altman 1991). At E20, activin A mRNA expression is highly consolidated in a laminar distribution between the cortical plate and layer IV. Later, at postnatal day 6, expression predominates in maturing neurons of layers III and II. Taken together, these observations suggest that, at a given stage of cortical development, expression of activin A prefigures the subsequent maturation of cortical neurons within that lamina.

At all developmental stages, activin A expression decreases abruptly ventral to the rhinal fissure, suggesting a function unique to neocortical development. In addition, fetal and early postnatal expression of activin A is notable for its sharp restriction to postmitotic neostriatal compartments, where expression is in patches with low or undetectable levels in the striatal matrix. Given the precise organization of cortical inputs to the striatum, characterized by afferents from specific frontal regions (Alexander et al. 1986) and segregated input to the patch and matrix compartments (Gerfen 1992), activin A may play a role in the development of topographically ordered sets of connections between neocortical and neostriatal compartments.

Recent studies have suggested that activin A can promote the differentiation of specific neuronal phenotypes or the elaboration of particular neuropeptides and neurotransmitters (Fann and Patterson 1994; Daadi et al. 1998; Darland and Nishi 1998; Ai et al. 1999; Galter et al. 1999; MacConell et al. 1999; Davis et al. 2000). Activin A may also have neurotrophic properties during development and following neuronal injury (Ameerun et al. 1996; Iwahori et al. 1997; Trudeau et al. 1997; Hughes et al. 1999; Tretter et al. 2000). Altogether, these data suggest that activin A is expressed in specific temporal and spatial patterns in response to specific signals to modulate the initial stages of neuronal differentiation early in development, and, at a later time, the neuronal response to synaptic activation.

BDNF. The neurotrophin brain-derived neurotrophic factor (BDNF), which promotes the activity-dependent survival of cortical neurons (Ghosh et al. 1994a), has also been identified as a nIEG that is regulated by excitatory synaptic activity in the adult brain (Lauterborn et al. 1996; Castren et al. 1998). In contrast to activin in the developing brain, BDNF mRNA has been localized to postnatal visual cortex in rodents shortly following eye-opening (Castren et al. 1992). In binocular animals, it has been found that BDNF is dynamically regulated in layer IV during the formation of ocular dominance columns (Lein et al. 2000) and, in the lateral geniculate nucleus, the origin of thalamocortical axons. During the formation of ocular dominance columns, BDNF is believed to function in activity-dependent axonal and dendritic remodeling. Thus, these growth factor nIEGs, activin A and BDNF, may function at specific periods of cortical development to promote neuronal differentiation, survival, or synaptic plasticity.

4.2
Extracellular Matrix and Signaling Molecules: Arcadlin, tPA, and Narp

Arcadlin. Arcadlin (*A*ctivity-*r*egulated *cad*herin-*li*ke protei*n*) is a novel nIEG, whose expression is tightly regulated by excitatory synaptic activity in adult hippocampus and cerebral cortex (Yamagata et al. 1999). Arcadlin is a member of the protocadherin family of calcium-dependent adhesion molecules. Cadherins are Ca^{2+}-dependent cell-cell adhesion molecules (CAMs) that regulate morphogenesis of many organs (Redies 1996) and encode integral membrane glycoproteins comprised of large extracellular, single transmembrane, and cytoplasmic domains. The extracellular domain contains functional repeats that each includes a Ca^{2+}-binding site. Classical cadherins have a conserved tripeptide (Blaschuk et al. 1990; Doherty et al. 1991) in the first extracellular subdomain, which in concert with surrounding sequence determines the homophilic binding specificity (Nose et al. 1990).

Protocadherins in general (Sano et al. 1993), and Arcadlin in particular, differ from classical cadherins in the sequence composition of the cytoplasmic domain. Cadherins exert downstream effects through interactions with the catenin cytoskeletal proteins at conserved residues in the carboxy tail (Ozawa and Kemler 1992). This association is believed to mediate the attachment of cadherins to the actin network and signaling through the wnt/wingless pathway (McCrea et al. 1991; Mathur et al. 1994; Stappert and Kemler 1994). Neither Arcadlin nor the known protocadherins share homology with the aforementioned cadherin cytoplasmic domains, suggesting that Arcadlin and other protocadherins may interact with a novel class of proteins and signaling molecules.

During early brain development, borders of cadherin expression in the neuroepithelium often coincide with segmental boundaries (Redies 1996). Later, specific subsets of differentiating neurons express specific cadherin isoforms (Ganzler 1995). The delineation of cortical and subcortical structures is promoted by the differential adhesiveness of early differentiating neurons, which express specific cadherins and aggregate homotypically. With later development, structures that originate from different neuromeres, expressing the same cadherin, become connected to each other through fiber tracts to form functional neural circuits subserving different sensory modalities (Redies and Takeichi 1993). Studies of N-cadherin have demonstrated that cytoplasmic sequences distinct from those involved in catenin interactions mediate axonal outgrowth in retinal ganglion cells (Riehl et al. 1996). This suggests that cadherins may have multiple functions in the developing brain. They may promote aggregation and cell sorting during early development, and neurite outgrowth and axonal pathfinding at later stages.

Arcadlin is highly expressed during late fetal and early postnatal development in a subset of structures integrating auditory, visual, and limbic pathways which become functionally connected at this stage (Yamagata et al. 1999). For example, in the auditory system, high levels of Arcadlin expression occur in the inferior colliculus, the medial geniculate of the thalamus and the auditory cortex, all of which are connected to one another to form a functional circuit. In rats, at postnatal day 1, when thalamocortical connections are actively being made, Arcadlin mRNA is simultaneously expressed in the medial geniculate of the thalamus and its target auditory cortex. This suggests a role for Arcadlin in establishing auditory synaptic contacts in a homophilic manner. Structures involved in the developing limbic system, including hippocampus, habenula, anterodorsal and anteroventral limbic thalamic nuclei, temporal and entorhinal cortices, and the amygdala, also express high levels of Arcadlin. Finally, in the visual system, targets of retinal innervation, including the suprachiasmatic nucleus (via the retinohypothalamic tract), the superior colliculus, and the ventral lateral geniculate nucleus, all express high levels of Arcadlin mRNA. Interestingly, Arcadlin expression is distinctly absent in developing fetal structures subserving the somatosensory system, notably the large ventrolateral and ventrobasal thalamic nuclei and the somatosensory cortex. Perhaps another protocadherin or cadherin functions separately in the establishment of somatosensory circuitry.

As described previously, there is a substantial body of evidence indicating that spontaneous neural activity occurs during early development, prior to sensory experience, and is required for the proper patterning of connections (Katz and Shatz 1996). It is not yet fully known whether the formation of the auditory, limbic and visual connections characterized by Arcadlin expression is activity-dependent. However, given the precedent from studies in the visual system, in which spontaneous retinal activity patterns subcortical structures such as the lateral geniculate (Wong et al. 1993; Mooney et al. 1996), Arcadlin's distribution raises the interesting possibility that activity-dependent expression of this nIEG promotes synaptic connections within components of a sensory circuit during development. It is also reasonable to hypothesize that similar activity-dependent processes may be occurring in other circuits subserving other sensory modalities. N-Cadherin has been localized by electron microscopy to the synaptic cleft, specifically to adherence junctions that border the neurotransmitter release zone (Uchida et al. 1996). Analogously, Arcadlin is localized to synapses in adult cortical neurons (Yamagata et al. 1999). The homophilic binding properties of Arcadlin may provide a molecular basis for activity-dependent adhesive interactions between opposing synaptic membranes, and which can thus modify synaptic structure in the adult brain. Similar mechanisms may operate during cortical and subcortical synapse formation and elimination.

tPA. Tissue plasminogen activator (tPA) is a serine protease expressed in neurons of the hippocampus and cerebral cortex (Davies et al. 1998), which may play a role in structural changes associated with activity-dependent plasticity. The substrate of tPA is plasminogen, a lipid synthesized in neurons. The conversion by activated tPA of plasminogen to the broad-spectrum protease plasmin allows the cleavage of a wide range of extracellular proteins including laminin and NCAM. Expression of tPA has been found to be rapidly induced following plastic paradigms, such as seizures and LTP (Qian et al. 1993), as well as in cerebellar Purkinje cells by the learning of complex motor tasks (Seeds et al. 1995). tPA is localized to presynaptic boutons and is released in response to synaptic activity (Gualandris et al. 1996; Muller and Griesinger 1998).

A role for extracellular proteolysis in activity-dependent synaptic plasticity is suggested by the findings that there is a decreased late phase of hippocampal LTP both in tPA null mice and following administration of tPA inhibitors (Huang et al. 1996; Baranes et al. 1998). tPA null mice have also shown a deficit in specific learning tasks, in particular two-way active avoidance. A second action of tPA, which may be distinct from and not involve its proteolytic activity, is believed to be mediated through binding of tPA to its receptor, the low-density lipoprotein receptor-related protein (LRP), a multifunctional endocytic receptor that is abundantly expressed in neurons (Zhuo et al. 2000). In this study, tPA promoted late-phase LTP by binding to LRP and potentiating the activity of cAMP-dependent protein kinase in hippocampal neurons.

Narp. Narp (*N*euronal *a*ctivity *r*egulated *p*entraxin) is a member of the long pentraxin family (Goodman et al. 1996). Members of this family of proteins include serum amyloid protein and C-reactive protein. Pentraxins share the properties of being multimeric secreted proteins (Gewurz et al. 1995) which, in the presence of Ca^{2+}, can bind to glycoproteins and glycolipids (Zahedi 1996; Saito et al. 1997; Wang and Sui 1999; Valentin and Lambeau 2000). Narp mRNA is enriched in neurons and is rapidly induced in hippocampus and cortex by depolarizing stimuli. Narp expression is developmentally regulated in cerebral cortex with peak expression, as for other nIEGs (i.e., transcription factors and effector IEGs; Kaufmann et al. 1994), during the third postnatal week of development (Tsui et al. 1996). Thus Narp, like all other known members of the pentraxin family, as a secreted protein may have the capacity to bind to glycoproteins in the extracellular space and impact on synaptic physiology and transmission. Lectin-binding sites, detected using plant lectins such as concanavalin A or pea lectin, are prominently expressed in the developing cortex and have been shown to be associated with discrete populations of neurons (Waters et al. 1990; Luth et al. 1992). Because Narp is dynamically regulated by physiological activity, it may function in immediate modifications of extracellular component proteins and lipids that function in synaptic physiology and transmission.

Recent studies have determined that Narp is enriched in a subpopulation of excitatory synapses in vivo and in vitro (O'Brien et al. 1999). In vitro, Narp is localized to a subset of axodendritic shafts on the aspiny spinal cord and hippocampal synapses between glutamatergic pyramidal neurons and GABAergic interneurons. Interestingly, in neuronal cultures, Narp is selectively transported to specific regions of the neuron, depending on whether it is an inhibitory or excitatory neuron. Narp localizes to dendrites in the case of inhibitory neurons, and to axonal terminals in the case of glutamatergic neurons. Significantly, Narp has also been shown to mediate the extracellular clustering of a specific subset of AMPA receptors, Glu R1, Glu R2 and Glu R3, which mediate fast excitatory synaptic transmission at a subset of glutamatergic synapses. Parallels have been drawn between the clustering action of Narp and that of agrin (Cohen et al. 1997), which signals transsynaptically through the tyrosine kinase MuSK and rapsyn to mediate clustering of the acetylcholine receptor. By electron microscopy, Narp is present at both pre- and postsynaptic sites of spiny and aspiny synapses. Because Narp is exquisitely regulated by excitatory synaptic activity in adult brain neurons, this nIEG may play a critical role in linking activity with the ontogeny/development of excitatory synapses. Finally, overexpression of Narp in cultured neurons can increase the number of presynaptic terminals, suggesting that Narp may participate in synaptogenesis.

4.3
Cytoskeletal Molecules: Arc

Arc (*A*ctivity *r*egulated *c*ytoskeleton-associated protein) encodes an IEG product with homology to spectrin which is associated with the actin cytoskeleton (Lyford et al. 1995). Arc mRNA is markedly enriched in adult brain, and its expression is rapidly and strongly induced in hippocampus and cortex by depolarizing stimuli such as seizures and LTP. Similarly, in dopamine-mediated paradigms of synaptic activation, Arc expression is induced rapidly in striatum (Fosnaugh et al. 1995; Berke et al. 1998; Tan et al. 2000). Arc is dynamically regulated in the mature brain in the CA1 region of the hippocampus and in the cerebral cortex, in response to natural stimuli such as exposure to a novel environment (Guzowski et al. 1999). Inhibition of Arc translation in in vivo LTP has been shown to inhibit the maintenance of this plastic change, as well as impair memory consolidation (Guzowski et al. 2000).

Arc is distinct from other known nIEGs in that both its mRNA and protein are enriched in dendrites of neurons (Steward et al. 1998; Wallace et al. 1998). Interestingly, Arc is the only nIEG mRNA to be transported out into the dendritic compartment. Other genes whose mRNA is selectively transported to the dendritic compartment include the alpha-subunit of calmodulin-dependent protein kinase II (CAM II) and high molecular weight microtubule-associate protein (MAP-2; Steward 1997). Although the function of Arc at the synapse has not been established, it can interact with polymerized actin, suggesting a role in activity-dependent modifications of the dendritic cytoskeletal structure. Moreover, following synaptic activation, Arc mRNA is selectively targeted to activated dendritic sites where Arc protein is synthesized at polyribosomal complexes associated with these postsynaptic sites (Steward et al. 1998). Inhibition of protein synthesis does not disrupt the selective targeting of Arc mRNA to these dendritic sites, however, suggesting that the transcript localization signals reside in the Arc mRNA sequence.

Arc mRNA is developmentally regulated in the cortex with expression first detected at postnatal day 8, and a peak at postnatal day 21. Arc's ontogeny, as that of transcription factor nIEGs (Worley et al. 1990a), parallels that of NMDA and environmentally dependent cortical plasticity (Fox and Daw 1993; Williams et al. 1993; Fox and Zahs 1994). In the adult, Arc mRNA basal expression is localized to layers II/III and V/VI of cortex, which originate cortical efferents, and to granule cells of the dentate gyrus and pyramidal neurons of hippocampus. While the temporal and spatial patterns of Arc expression suggest a role for this nIEG in activity-dependent neuronal plasticity, its association with the neuronal cytoskeleton indicates a role in structural alterations that may be associated with changes in neuronal connectivity.

4.4 Signaling Molecules: Rheb and COX-2

Rheb. Rheb (*R*as *h*omologue *e*nriched in *b*rain) is a novel member of the Ras family of small GTP-binding proteins (Yamagata et al. 1994). Studies of Ras family members suggest that Ras proteins function to regulate the activity of specific biochemical pathways (Reuther and Der 2000). Different members of the Ras family regulate diverse functions such as growth factor dependent signaling, cell polarity, and vesicle sorting and transport. Rheb is closely homologous to H-Ras, particularly in the G2 effector domain that is involved in GTP binding and coupling of H-Ras to its GTPase activating protein, as well as to the serine/threonine kinase Raf-1 (Boguski and McCormick 1993), sharing with the latter its ability to transform cells. Ras proteins control a diversity of cellular functions through interactions with specific target proteins. Rheb interacts with Raf-1 kinase, an essential component of the H-Ras signaling pathway (Yee and Worley 1997). In direct contrast to H-Ras, however, the interaction of Rheb with Raf-1 is potentiated by growth factors and agents that increase cyclic AMP (cAMP) levels. Protein kinase A dependent phosphorylation of serine 43, within the regulatory domain of Raf-1, reciprocally potentiates its interaction with Rheb and decreases its interaction with H-Ras. These data suggest a mechanism in which Rheb functions in concert with H-Ras to dynamically integrate cAMP and growth factor signaling.

Unlike other Ras family members, Rheb is regulated as an IEG in the brain, being induced in granule cells of the hippocampus by both depolarizing stimuli and by tetanic synaptic stimuli associated with LTP. During brain development, Rheb mRNA is prominently and widely expressed in developing (prenatal) cortical plate, as well as in postnatal developing and adult cortex. Although the function of Rheb in neurons remains to be defined, its rapid induction by growth factors and neurotransmitters suggests a role in activity-dependent growth and differentiation of neurons in the developing cortex.

COX-2. The cyclooxygenases (COXs) or prostaglandin synthases are the rate-limiting enzymes in the production of prostaglandins and prostanoids from arachidonic acid (Smith et al. 1991). These enzymes are inhibited by non-steroidal antiinflammatory drugs (NSAIDs), such as aspirin, indomethacin and ibuprofen. There are two isoforms of COX: COX-1 is constitutively expressed in brain, at very low levels, predominantly in astrocytes and microglia (Kaufmann et al. 1997; Yermakova et al. 1999). COX-2, the inducible isoform of cyclooxygenase (Yamagata et al. 1993; Herschman 1994), is expressed selectively in neocortical layers II/III and V, as well as in layer II of the entorhinal and piriform cortices, and the hippocampus. In the latter, COX-2 is expressed predominantly in the dentate gyrus, hilus, CA3, and subiculum (Breder et al. 1995; Kaufmann et al. 1996). COX-2 is also highly expressed in the deep nuclei of the amygdala and at a lower level in the hypothalamus. COX-2 expression increases dramatically and transiently in response to glutamatergic neurotransmission (Yamagata et al. 1993).

Within a particular region, basal COX-2 immunoreactivity localizes regionally in a pattern that resembles that of transcription factor nIEGs (Breder et al. 1995; Kaufmann et al. 1996). Within individual neurons, COX-2 is perikaryal, with enrichment in perinuclear membranes and dendrites. COX-2 immunoreactivity colocalizes morphologically to dendritic spines of pyramidal and other excitatory neurons in cerebral cortex and limbic regions (Kaufmann et al. 1996). Dendritic spines are postsynaptic sites of NMDA receptor-mediated neurotransmission, raising the interesting possibility that production of prostaglandins at synaptically active dendritic spines may directly modulate synaptic morphology and function. During development, COX-2 is only expressed significantly postnatally, with mRNA first detected in hippocampus at day 13 and, subsequently, in cerebral cortex by day 21. High basal levels are maintained in these selected neuronal circuits, but can be rapidly downregulated in the presence of NMDA-receptor antagonists or interruption of afferent excitatory input (Yamagata et al. 1993).

The mechanism by which products of COX-2 metabolism of arachidonic acid contribute to synaptic plasticity in the adult and developing brain is beginning to be understood. Activity-dependent changes in neuronal connectivity require activation of glutamate receptors, in particular of the NMDA subtype. Intracellular signaling cascades can couple the activation of NMDA receptors at the neuronal membrane with induction of nIEG expression in the cell nucleus. Following NMDA activation, intracellular concentration of Ca^{2+} increases leading to the activation of phospholipase A2 (PLA2) and release of arachidonic acid (Lerea et al. 1995). Cyclooxygenase activity, resulting in the generation of PGF2alpha, leads to the induction of the transcription factor IEG c-fos (Lerea et al. 1997). PGF2alpha has been found to be necessary, but not sufficient for, NMDA receptor induced expression of the IEG c-fos (Lerea et al. 1997). Synthesis of PGF2alpha may occur locally in the neuron's dendritic spines, where NMDA receptors and COX-2 are co-localized. The mechanism by which PGF2alpha regulates the downstream expression of nIEGs is not known, but presumably is mediated via the PGF2alpha receptor, a seven-transmembrane domain G-protein linked receptor that is enriched in brain (Hirata et al. 1995; Sakamoto et al. 1995).

Clues to the biochemical and phenotypic changes that occur as a result of COX activity in neurons can be gleaned from studies in non-neuronal cells. Proliferative responses induced by epidermal growth factor (EGF) are mediated by the release of arachidonic acid by PLA2, from plasma membrane phospholipids, and subsequent induction of a class of transcription factor IEGs (Clark and Dunlop 1991). Inhibition of leukotriene and prostaglandin production (Handler et al. 1990; Sellmayer et al. 1991) can inhibit this mitogenic response to EGF. The metabolism of arachidonic acid in rat fibroblasts, following EGF stimulation, causes a dramatic rounding up of the cells. In this system, prostaglandin synthesis induces the breakdown of stress actin fibers whereas leukotrienes induce actin polymerization in the cellular cortex (Peppelenbosch et al. 1993). In an analogous neuronal system, stimulation of pheochromocytoma PC12 cells by nerve growth factor (NGF) typically results in outgrowth of neurites. This type of NGF stimulation has been shown to increase arachidonate metabolism and prostaglandin/prostanoid production, and neurite differentiation can be blocked by inhibitors of PLA2 and lipoxygenase metabolism (DeGeorge et al. 1988). Thus, one potential mechanism of action of COX-2 metabolites in neuronal plasticity may be in mediating alterations in cytoskeletal synaptic structure.

4.5
Anchoring/Coupling Proteins: Homer

Homer 1a is a neuronal activity-modulated member of a novel class of anchoring proteins derived from three genes termed Homer 1–3 (Brakeman et al. 1997; Xiao et al. 1998; Tu et al. 1999). Homer 1a is developmentally regulated, with peak expression in the rat forebrain between the third and the fifth postnatal weeks. During this period of peak developmental expression, Homer 1a mRNA is markedly induced in cerebral cortex of dark-reared rats within

30 min of their first visual experience. This experimental paradigm has been previously used in rodents, cats and monkeys, to examine the role of visual activity in the normal development of the visual cortex. In another rat model of visual system plasticity, in which monocular deprivation induced by tetrodotoxin blocks retinal activity and afferent input to the visual cortex, Homer 1a mRNA is rapidly downregulated in the contralateral visual cortex. These observations indicate that, as with transcription factor nIEGs, the developmental expression of Homer in visual cortex is regulated by natural synaptic activity (Brakeman et al. 1997). In the adult, Homer 1a mRNA is rapidly induced in hippocampus by NMDA-dependent synaptic stimuli that induce LTP, and in the striatum by dopaminergic activity.

Homer proteins contain an N-terminal EVH1 domain homologous to the Ena/VASP family of proteins. The EVH1 domain mediates the binding of Homer proteins to proline-rich sequences in the intracellular C-terminus of group I metabotropic glutamate receptors (mGluRs; Xiao et al. 1998; Tu et al. 1999), as well as to sequences in the inositol triphosphate receptor (IP3R), ryanodine receptors, and Shank family proteins (Beneken et al. 2000; Xiao et al. 2000). Homer proteins, with the exception of the nIEG Homer 1a splice variant, also contain a C-terminal coiled-coil domain that mediates self-multimerization. These Homer proteins are therefore able to self-associate and cross-link with interacting proteins (Xiao et al. 2000). The Homer splice forms are highly conserved in the region that binds mGluR5, while they are divergent in their carboxyl regions. In fact, the Homer 1a splice variant does not possess the C-terminal leucine zipper motif, and, for this reason, is able to function as a dominant negative form of the constitutively expressed longer forms of Homer proteins. Increased expression of Homer 1a at the synapse, following synaptic activation, may disrupt coupling between interacting proteins that are mediated by long Homer forms (Kammermeier et al. 2000).

Metabotropic receptors (mGluRs) are essential for the establishment of long-term synaptic plasticity in models of learning and memory (Aiba et al. 1994; Bortolotto et al. 1999), as well as during neuronal development (Catania et al. 1994). Binding of Homer proteins to this important receptor may represent a mechanism of activity-dependent modification of synaptic physiology. It is hypothesized that, by binding mGluRs, Homer plays a regulatory role in the spatial targeting and/or signaling properties of this type of glutamate receptor. In support of the first mechanism, recent studies have shown that the localization of mGluR1a and mGluR5 to dendritic perisynaptic and extrasynaptic sites (Baude et al. 1993; van den Pol 1994; van den Pol et al. 1995) seems to be dependent on the type of Homer 1 isoform that is expressed (Ango et al. 2000). It is interesting to note that glutamate-induced phosphoinositol turnover in the visual cortex peaks during the postnatal critical period and can be artificially maintained by rearing animals in the dark, suggesting an activity-dependent modulation of this response (Dudek and Bear 1989). Accordingly, Homer 1a may play a role in activity-dependent targeting and signaling properties of mGluRs during neocortical development.

5
Conclusions

A compelling picture is emerging from this work regarding the general function of IEGs in the regulation of neuronal structure and physiology (Fig. 3). Both transcription factor and effector neuronal IEGs participate in an

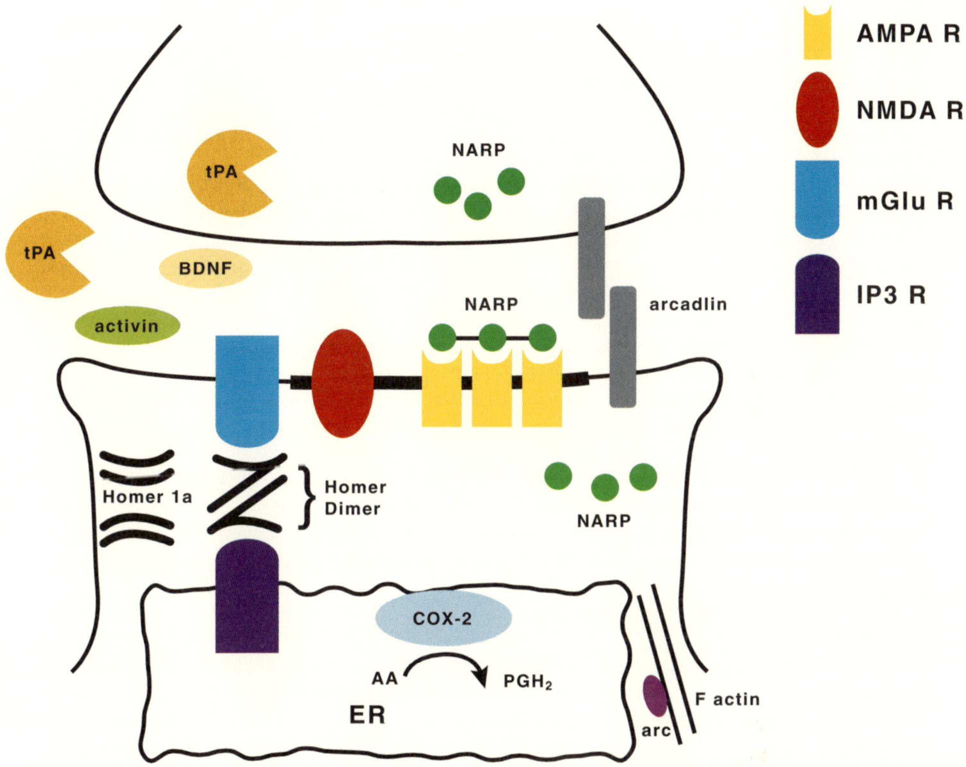

Fig. 3. Model of localization of effector neuronal immediate early genes. *Activin A*, a member of the TGFβ family of secreted growth factors, may be synthesized and secreted both pre- and postsynaptically as has been shown with *BDNF*. *Narp*, a member of the long pentraxin family, is present both pre- and postsynaptically, and when secreted will oligomerize and cluster GluR1–3 subunits of *AMPA* receptors. *tPA* has been found in presynaptic boutons and can be released into the synaptic cleft in response to synaptic activity and act as a protease on a range of extracellular proteins. *Arcadlin*, similarly to classical cadherins, is postulated to mediate adhesion between pre- and postsynaptic sites in a Ca^{2+}-dependent manner. The constitutively expressed anchoring *Homer* proteins are dimers that can create a physical link between subunits of the metabotropic glutamate receptors (*mGlu R*) and the inositol (1,4,5)P3 receptor (*IP3 R*). The IEG *Homer 1a* does not contain the leucine zipper sequence that allows dimerization, and therefore can compete with constitutively expressed Homer proteins for *mGlu R* and *IP3 R* and disrupt these protein-protein interactions. *Arc* is a cytoskeletal protein that can interact with *F-actin*, and its mRNA is present in dendrites. *COX-2*, the rate-limiting enzyme in the formation of the prostaglandin PGH_2 from arachidonic acid (*AA*), is localized on the lumenal surface of the endoplasmic reticulum (*ER*)

activity-dependent process in which signaling and metabolic pathways, protein-protein interactions, and cytoskeletal elements can be rapidly modified in response to excitatory synaptic activity. Consequently, the identification and functional characterization of these classes of neuronal IEGs will yield insight into the mechanisms by which excitatory activity leads to long-term changes in synaptic and cellular properties.

References

Ai X, Cappuzzello J, Hall AK (1999) Activin and bone morphogenetic proteins induce calcitonin gene-related peptide in embryonic sensory neurons in vitro. Mol Cell Neurosci 14:506–518

Aiba A, Chen C, Herrup K, Rosenmund C, Stevens CF, Tonegawa S (1994) Reduced hippocampal long-term potentiation and context-specific deficit in associative learning in mGluR1 mutant mice. Cell 79:365–375

Akbarian S, Bunney WE, Potkin SG (1993) Altered distribution of nicotinamide-adenine dinucleotide phosphate-diaphorase cells in frontal lobe of schizophrenics implies disturbances of cortical development. Gen Psychiatry 50:169–177

Alexander GE, DeLong MR, Strick PL (1986) Parallel organization of functionally segregated circuits linking basal ganglia and cortex. Annu Rev Neurosci 9:357–381

Ameerun RF, de Winter JP, van den Eijnden-van Raaij AJ, den Hertog J, de Laat SW, Tertoolen LG (1996) Activin and basic fibroblast growth factor regulate neurogenesis of murine embryonal carcinoma cells. Cell Growth Differ 7:1679–1688

Anderson SA, Volk DW, Lewis DA (1996) Increased density of microtubule associated protein 2-immunoreactive neurons in the prefrontal white matter of schizophrenic subjects. Schizophr Res 19:111–119

Andreasson K, Worley P (1995) Induction of β-A activin expression by synaptic activity and during neocortical development. Neuroscience 69:781–796

Ango F, Pin JP, Tu JC, Xiao B, Worley PF, Bockaert J, Fagni L (2000) Dendritic and axonal targeting of type 5 metabotropic glutamate receptor is regulated by homer1 proteins and neuronal excitation. J Neurosci 20:8710–8716

Artola A, Singer W (1987) Long-term potentiation and NMDA receptors in rat visual cortex. Nature 330:649–652

Bading H, Ginty DD, Greenberg ME (1993) Regulation of gene expression in hippocampal neurons by distinct calcium signaling pathways. Science 260:181–186

Baranes D, Lederfein D, Huang YY, Chen M, Bailey CH, Kandel E (1998) Tissue plasminogen activator contributes to the late phase of LTP and to synaptic growth in the hippocampal mossy fiber pathway. Neuron 21:813–825

Baude A, Nusser Z, Roberts JD, Mulvihill E, McIlhinney RA, Somogyi P (1993) The metabotropic glutamate receptor (mGluR1 alpha) is concentrated at perisynaptic membrane of neuronal subpopulations as detected by immunogold reaction. Neuron 11:771–787

Bayer SA, Altman J (1991) Neocortical development. Raven Press, New York

Bear MF, Kleinschmidt A, Gu Q, Singer W (1990) Disruption of experience-dependent synaptic modifications in striate cortex by infusion of an NMDA receptor antagonist. J Neurosci 10: 909–925

Beneken J, Tu JC, Xiao B, Nuriya M, Yuan JP, Worley PF, Leahy DJ (2000) Structure of the Homer EVH1 domain-peptide complex reveals a new twist in polyproline recognition. Neuron 26: 143–154

Berke JD, Paletzki RF, Aronson GJ, Hyman SE, Gerfen CR (1998) A complex program of striatal gene expression induced by dopaminergic stimulation. J Neurosci 18:5301–5310

Blanton MG, LoTurco JJ, Kriegstein AR (1990) Endogenous neurotransmitter activates N-methyl-D-aspartate receptors on differentiating neurons in embryonic cortex. Proc Natl Acad Sci USA 87:8027–8030

Blaschuk OW, Sullivan DS, Pouliot Y (1990) Identification of a cadherin cell adhesion recognition sequence. Dev Biol 139:227–229
Bliss TVP, Collingridge GL (1993) A synaptic model of memory: long-term potentiation in the hippocampus. Nature 361:31–39
Boguski MS, McCormick F (1993) Proteins regulating Ras and its relatives. Nature 366:643–654
Bortolotto ZA, Fitzjohn SM, Collingridge GL (1999) Roles of metabotropic glutamate receptors in LTP and LTD in the hippocampus. Curr Opin Neurobiol 9:299–304
Brakeman PR, Lanahan AA, O'Brien R, Roche K, Barnes CA, Huganir RL, Worley PF (1997) Homer: a protein that selectively binds metabotropic glutamate receptors. Nature 386:284–288
Breder CD, Dewitt D, Krang RP (1995) Characterization of inducible cyclooxygenase in rat brain. J Comp Neurol 355:296–315
Castren E, Zafra F, Thoenen H, Lindholm D (1992) Light regulates expression of brain-derived neurotrophic factor mRNA in rat visual cortex. Proc Natl Acad Sci USA 89:9444–9448
Castren E, Berninger B, Leingartner A, Lindholm D (1998) Regulation of brain-derived neurotrophic factor mRNA levels in hippocampus by neuronal activity. Prog Brain Res 117:57–64
Catania MV, Landwehrmeyer GB, Testa CM, Standaert DG, Penney JB, Young AB (1994) Metabotropic glutamate receptors are differentially regulated during development. Neuroscience 61:481–495
Chaudhuri A, Cynader MS (1993) Activity-dependent expression of the transcription factor Zif268 reveals ocular dominance columns in monkey visual cortex. Brain Res 605:349–353
Clark S, Dunlop M (1991) Modulation of phospholipase A2 activity by epidermal growth factor in CHO cells transfected with human EGF receptor. Biochem J 274:715–721
Cochran BH, Keffel AC, Stiles CD (1983) Molecular cloning of gene sequences regulated by platelet-derived growth factor. Cell 33:939–947
Cohen NA, Kaufmann WE, Worley PF, Rupp F (1997) Expression of agrin in the developing and adult rat brain. Neuroscience 76:581–596
Cole A, Saffen D, Baraban J, Worley P (1989) Rapid increase of an immediate early gene mRNA in hippocampal neurons by synaptic NMDA receptor activation. Nature 340:474–476
Constantine-Paton M, Cline HT, Debski E (1990) Patterned activity, synaptic convergence, and the NMDA receptor in developing visual pathways. Annu Rev Neurosci 13:129–154
Crair MC, Malenka RC (1995) A critical period for long-term potentiation at thalamocortical synapses. Nature 375:325–328
Curran T, Franza BR (1988) Fos and Jun: the AP-1 connection. Cell 55:395–397
Daadi M, Arcellana-Panlilio MY, Weiss S (1998) Activin co-operates with fibroblast growth factor 2 to regulate tyrosine hydroxylase expression in the basal forebrain ventricular zone progenitors. Neuroscience 86:867–880
Darland DC, Nishi R (1998) Activin A and follistatin influence expression of somatostatin in the ciliary ganglion in vivo. Dev Biol 202:293–303
Davies BJ, Pickard BS, Steel M, Morris RG, Lathe R (1998) Serine proteases in rodent hippocampus. J Biol Chem 273:23004–23011
Davis AA, Matzuk MM, Reh TA (2000) Activin A promotes progenitor differentiation into photoreceptors in rodent retina. Mol Cell Neurosci 15:11–21
DeGeorge JJ, Walenga R, Carbonetto S (1988) Nerve growth factor rapidly stimulates arachidonate metabolism in PC12 cells: potential involvement in nerve fiber growth. J Neurosci Res 21:323–332
Dobbing J, Sands J (1979) Comparative aspects of the brain growth spurt. Early Hum Dev 3:79–83
Dobbing J, Smart JL (1974) Vulnerability of developing brain and behavior. Br Med Bull 30: 164–168
Doherty P, Rowett LH, Moore SE, Mann DE, Walsh FS (1991) Neurite outgrowth in response to transfected N-CAM and N-cadherin reveals fundamental differences in neuronal responsiveness to CAMs. Neuron 6:247–258
Dudek SM, Bear MF (1989) A biochemical correlate of the critical period for synaptic modification in the visual cortex. Science 246:673–675

Fann MJ, Patterson PH (1994) Neuropoietic cytokines and activin A differentially regulate the phenotype of cultured sympathetic neurons. Proc Natl Acad Sci USA 91:43–47

Farmer SF, Harrison LM, Ingram DA, Stephens JA (1991) Plasticity of the central motor pathways in children with hemiplegic cerebral palsy. Neurology 41:1505–1510

Flor PJ, Gomeza J, Tones MA, Kuhn R, Pin JP, Knopfel T (1996) The C-terminal domain of the mglur1 metabotropic glutamate receptor affects sensitivity to agonists. J Neurochem 67:58–63

Fosnaugh JS, Bhat RV, Yamagata K, Worley PF, Baraban JM (1995) Activation of arc, a putative "effector" immediate early gene, by cocaine in rat brain. J Neurochem 64:2377–2380

Fox K, Daw NW (1993) Do NMDA receptors have a critical function in visual cortical plasticity? Trends Neurosci 16:116–122

Fox K, Zahs K (1994) Critical period control in sensory cortex. Curr Opin Neurobiol 4:112–119

Fox K, Sato H, Daw NW (1989) The location and function of NMDA receptors in cat and kitten visual cortex. J Neurosci 9:2443–2454

Galter D, Bottner M, Unsicker K (1999) Developmental regulation of the serotonergic transmitter phenotype in rostral and caudal raphe neurons by transforming growth factor-betas. J Neurosci Res 56:531–538

Ganzler SIaRC (1995) R-Cadherin expression during nucleus formation in chicken forebrain neuromeres. J Neurosci 15:4157–4172

Gerfen CR (1992) The neostriatal mosaic: multiple levels of compartmental organization. Trends Neurosci 15:133–139

Gewurz H, Zhang XH, Lint TF (1995) Structure and function of the pentraxins. Curr Opin Immunol 7:54–64

Ghosh A, Shatz CJ (1992) Involvement of subplate neurons in the formation of ocular dominance columns. Science 255:1441–1443

Ghosh A, Carnahan J, Greenberg ME (1994a) Requirement for BDNF in activity-dependent survival of cortical neurons. Science 263:1618–1623

Ghosh A, Ginty DD, Bading H, Greenberg ME (1994b) Calcium regulation of gene expression in neuronal cells. J Neurobiol 25:294–303

Goodman AR, Cardozo T, Abagyan R, Altmeyer A, Wisniewski HG, Vilcek J (1996) Long pentraxins: an emerging group of proteins with diverse functions. Cytokine Growth Factor Rev 7:191–202

Goodman CS, Shatz CJ (1993) Developmental mechanisms that generate precise patterns of neuronal connectivity. Cell/Neuron 72(10):77–98

Greenough WT (1984) Structural correlates of information storage in the mammalian brain: a review and hypothesis. Trends Neurosci 7:229–233

Gruzellier J, Flor-Henry P (1979) Hemispheric asymmetries of function in psychopathology. Elsevier, North Holland, New York

Gualandris A, Jones TE, Strickland S, Tsirka SE (1996) Membrane depolarization induces calcium-dependent secretion of tissue plasminogen activator. J Neurosci 16:2220–2225

Guzowski JF, McNaughton BL, Barnes CA, Worley PF (1999) Environment-specific expression of the immediate-early gene Arc in hippocampal neuronal ensembles. Nat Neurosci 2:1120–1124

Guzowski JF, Lyford GL, Stevenson GD, Houston FP, McGaugh JL, Worley PF, Barnes CA (2000) Inhibition of activity-dependent arc protein expression in the rat hippocampus impairs the maintenance of long-term potentiation and the consolidation of long-term memory. J Neurosci 20:3993–4001

Hagihara K, Tsumoto T, Sato H, Hata Y (1988) Actions of excitatory amino acid antagonists on geniculo-cortical transmission in the cat's visual cortex. Exp Brain Res 69:407–416

Handler JA, Danilowicz RM, Eling TE (1990) Mitogenic signaling by epidermal growth factor (EGF), but not platelet derived growth factor, requires arachidonic acid metabolism in BALB/c 3T3 cells. J Biol Chem 265:3669–3673

Herschman HR (1994) Regulation of prostaglandin synthase-1 and prostaglandin synthase-2. Cancer Metast Rev 13:241–256

Hirata M, Ushikubi F, Narumiya S (1995) Prostaglandin I receptor and prostaglandin D receptor. J Lipid Mediat Cell Signal 12:393–404

Horton JC, Howking DR (1996) An adult-like pattern of ocular dominance columns in striate cortex of newborn monkeys prior to visual experience. J Neurosci 16:1791–1807

Huang YY, Bach ME, Lipp HP, Zhuo M, Wolfer DP, Hawkins RD, Schoonjans L, Kandel ER, Godfraind JM, Mulligan R et al. (1996) Mice lacking the gene encoding tissue-type plasminogen activator show a selective interference with late-phase long-term potentiation in both Schaffer collateral and mossy fiber pathways. Proc Natl Acad Sci USA 93:8699–8704

Hubel DH, Wiesel TN (1970) The period of susceptibility to the physiological effects of unilateral eye closure in kittens. J Physiol 206:419–436

Hubel DH, Wiesel TN, LeVay S (1977) Plasticity of ocular dominance columns in monkey striate cortex. Philos Trans R Soc Lond B 278:377–409

Hughes PE, Alexi T, Williams CE, Clark RG, Gluckman PD (1999) Administration of recombinant human Activin-A has powerful neurotrophic effects on select striatal phenotypes in the quinolinic acid lesion model of Huntington's disease. Neuroscience 92:197–209

Hunt SP, Pini A, Evan G (1987) Induction of *c-fos*-like protein in spinal cord neurons following sensory stimulation. Nature 328:632–634

Huttenlocher PR (1991) Dendritic and synaptic pathology in mental retardation. Pediatr Neurol 7:79–85

Iwahori Y, Saito H, Torii K, Nishiyama N (1997) Activin exerts a neurotrophic effect on cultured hippocampal neurons. Brain Res 760:52–58

Kammermeier PJ, Xiao B, Tu JC, Worley PF, Ikeda SR (2000) Homer proteins regulate coupling of group I metabotropic glutamate receptors to N-type calcium and M-type potassium channels. J Neurosci 20:7238–7245

Katz LC, Shatz CJ (1996) Synaptic activity and the construction of cortical circuits. Science 274:1133–1138

Kaufmann WE (2002) Cortical histogenesis. In: Aminoff MJ, Daroff RB (eds) Encyclopedia of the neurological sciences. Section on neuroanatomy and clinical localization (Masdeu JC ed). Academic Press, New York (in press)

Kaufmann WE, Moser HW (2000) Dendritic anomalies in disorders associated with mental retardation. Cereb Cortex 10:981–991

Kaufmann WE, Worley PF (1999) The role of early neural activity in regulating immediate early gene expression in the cerebral cortex. MRDD Res Rev 5:41–50

Kaufmann WE, Yamagata K, Andreasson KI, Worley PF (1994) Rapid response genes as markers of cellular signaling during cortical histogenesis: their potential in understanding mental retardation. Int J Dev Neurosci 12:263–271

Kaufmann WE, Worley PF, Pegg J, Bremer M, Isakson P (1996) Cox-2, a synaptically induced enzyme, is expressed by excitatory neurons at postsynaptic sites in rat cerebral cortex. Proc Natl Acad Sci USA 93:2317–2321

Kaufmann WE, Andreasson KI, Isakson PC, Worley PF (1997) Cyclooxygenases and the central nervous system. Prostaglandins 54:601–624

Kirkwood A, Bear MF (1994) Homosynaptic long-term depression in the visual cortex. J Neurosci 14:3404–3412

Kirkwood A, Lee HK, Bear MF (1995) Co-regulation of long-term potentiation and experience-dependent synaptic plasticity in visual cortex by age and experience. Nature 375:328–331

Kleinschmidt A, Bear MF, Singer W (1987) Blockade of "NMDA" receptors disrupts experience-dependent plasticity of kitten striate cortex. Science 238:355–358

Lau LF, Nathans D (1991) The hormonal control regulation of gene trascription. In: Cohen P, Foulkes JG (eds) Molecular aspects of cell regulation. Elsevier, Amsterdam, pp 257–293

Lauterborn JC, Rivera S, Stinis CT, Hayes VY, Isackson PJ, Gall CM (1996) Differential effects of protein synthesis inhibition on the activity-dependent expression of BDNF transcripts: evidence for immediate-early gene responses from specific promoters. J Neurosci 16:7428–7436

Lein ES, Hohn A, Shatz CJ (2000) Dynamic regulation of BDNF and NT-3 expression during visual system development. J Comp Neurol 420:1–18

Lerea LS, Carlson NG, McNamara JO (1995) N-Methyl-D-aspartate receptors activate transcription of c-fos and NGFI-A by distinct phospholipase A2-requiring intracellular signaling pathways. Mol Pharmacol 47:1119–1125
Lerea LS, Carlson NG, Simonato M, Morrow J, Roberts JL McNamara JO (1997) Prostaglandin F 2alpha is required for NMDA receptor-mediated induction of c-fos mRNA in dentate gyrus neurons. J Neurosci 17:117–124
Linzer DIH, Nathans D (1983) Growth-related changes in specific mRNAs of cultured mouse cells. Proc Natl Acad Sci USA 80:4271–4275
LoTurco JJ, Kriegstein AR (1991) Clusters of coupled neuroblasts in embryonic neocortex. Science 252:563–566
Luth HJ, Fischer J, Celio MR (1992) Soybean lectin binding neurons in the visual cortex of the rat contains parvalbumin and are covered by glial nets. J Neurocytol 21:211–221
Lyford GL, Yamagata K, Kaufmann WE, Barnes CA, Sanders LK, Copeland NG, Gilbert DJ, Jenkins NA, Lanahan AA, Worley PF (1995) Arc, a growth factor and activity-regulated gene, encodes a novel cytoskeleton-associated protein that is enriched in neuronal dendrites. Neuron 14: 1–20
MacConell LA, Lawson MA, Mellon PL, Roberts VJ (1999) Activin A regulation of gonadotropin-releasing hormone synthesis and release in vitro. Neuroendocrinology 70:246–254
Maffei L, Galli-Resta L (1990) Correlation of the discharges of neighboring rat retinal ganglion cells during prenatal life. Proc Natl Acad Sci USA 87:2861–2864
Mather JP, Woodruff TK, Krummen LA (1992) Paracrine regulation of reproductive function by inhibin and activin. Proc Soc Exp Biol Med 201:1–15
Mather JP, Moore A, Li RH (1997) Activins, inhibins, and follistatins: further thoughts on a growing family of regulators. Proc Soc Exp Biol Med 215:209–222
Mathur M, Goodwing L, Cowin P (1994) Interactions of the cytoplasmic domain of the desmosomal cadherin Dsg1 with plakoglobin. J Biol Chem 269:14075–14080
McConnell SK (1988) Fates of visual cortical neurons in the ferret after isochronic and heterochronic transplantation. J Neurosci 8:945–974
McCrea PD, Turck CW, Gumbiner B (1991) A homolog of the armadillo protein in *Drosophila* (plakoglobin) associated with E-cadherin. Science 254:1359–1361
Meister M, Wong ROL, Baylor DA, Shatz CJ (1991) Synchronous bursts of action potentials in ganglion cells of the developing mammalian retina. Science 252:939–943
Miller KD, Chapman B, Stryker MP (1989) Visual responses in adult cat visual cortex depend on N-methyl-D-aspartate receptors. Proc Natl Acad Sci USA 86:5183–5187
Molliver ME, Krist DA (1975) The fine structural demonstration of monoaminergic synapses in immature rat neocortex. Neurosci Lett 1:305–310
Mooney R, Penn AA, Gallego R, Shatz CJ (1996) Thalamic relay of spontaneous retinal activity prior to vision. Neuron 17:863–874
Morgan J, Cohen D, Hempstead J, Curran T (1987) Mapping patterns of c-fos expression in the central nervous system after seizure. Science 237:192–197
Mrzljak L, Uylings HB, Van Eden CG, Judas M (1990) Neuronal development in human prefrontal cortex in prenatal and postnatal stages. Prog Brain Res 85:185–222
Muller CM, Griesinger CB (1998) Tissue plasminogen activator mediates reverse occlusion plasticity in visual cortex. Nat Neurosci 1:47–53
Nose A, Tsufi K, Takeichi M (1990) Localization of specificity determining sites in cadherin cell adhesion molecules. Cell 61:147–155
O'Brien RJ, Xu D, Petralia RS, Steward O, Huganir RL, Worley P (1999) Synaptic clustering of AMPA receptors by the extracellular immediate-early gene product Narp. Neuron 23:309–323
O'Donovan M, Sernagor E, Sholomenko G, Ho S, Antal M, Yee W (1992) Development of spinal motor networks in the chick embryo. J Exp Zool 261:261–273
Ono K, Yamano T, Shimada M (1991) Formation of an ipsilateral corticospinal tract after ablation of cerebral cortex in neonatal rat. Brain Dev 13:348–351
Ozawa M, Kemler R (1992) Molecular organization of the uvomorulin-catenin complex. J Cell Biol 116:989–996

Peppelenbosch MP, Tertoolen LG, Hage WJ, de Laat SW (1993) Epidermal growth factor-induced actin remodeling is regulated by 5-lipoxygenase and cyclooxygenase products. Cell 74: 565–575

Petit TL, LeBoutillier JC, Gregorio A, Libstug H (1988) The pattern of dendritic development in the cerebral cortex of the rat. Brain Res 469:209–219

Qian Z, Gilbert ME, Colicos MA, Kandel ER, Kuhl D (1993) Tissue-plasminogen activator is induced as an immediate-early gene during seizure, kindling and long-term potentiation. Nature 361:453–457

Quian Z, Gilbert ME, Colicos MA, Kandel ER, Kuhl D (1993) Tissue plasminogen activator is induced as an immediate-early gene during seizure, kindling and long-term potentiation. Nature 361:453–457

Raff MC, Barres BA, Burne JF, Coles HS, Ishizaki Y, Jacobson MD (1993) Programmed cell death and the control of cell survival: lessons from the nervous system. Science 262:695–700

Rakic P (1976) Prenatal genesis of connections subserving ocular dominance in the rhesus monkey. Nature 261:467–471

Redies CaTM (1996) Cadherins in the developing central nervous system: an adhesive code for segmental and functional subdivisions. Development 180:413–423

Redies C, Takeichi M (1993) Expression of N-cadherin during development of the mouse brain. Dev Dyn 197:26–39

Redies C, Engelhart K, Takeichi M (1993) Differential expression of N- and R-cadherin in functional neuronal systems and other structures of the developing chicken brain. J Comp Neurol 333:398–416

Reiter HO, Stryker MP (1988) Neural plasticity without postsynaptic action potentials: less-active inputs become dominant when kitten visual cortical cells are pharmacologically inhibited. Proc Natl Acad Sci USA 85:3623–3627

Reiter HO, Waitzman DM, Stryker MP (1986) Cortical activity blockade prevents ocular dominance plasticity in the kitten visual cortex. Exp Brain Res 65:182–188

Reuther GW, Der CJ (2000) The Ras branch of small GTPases: Ras family members don't fall far from the tree. Curr Opin Cell Biol 12:157–165

Riehl R, Johnson K, Bradley R, Grunwald GB, Cornel E, Lilienbaum A, Holt CE (1996) Cadherin function is required for axon outgrowth in retinal ganglion cells in vivo. Neuron 17:837–848

Rosen KM, McCormack MA, Villa-Komaroff L, Mower GD (1992) Brief visual experience induces immediate early gene expression in cat visual cortex. Proc Natl Acad Sci USA 89:5437–5441

Saffen DW, Cole AJ, Worley PF, Christy BA, Ryder K, Baraban JM (1988) Convulsant-induced increase in transcription factor messenger RNAs in rat brain. Proc Natl Acad Sci USA 85:7795–7799

Saito T, Hatada M, Iwanaga S, Kawabata S (1997) A newly identified horseshoe crab lectin with binding specificity to O-antigen of bacterial lipopolysaccharides. J Biol Chem 272: 30703–30708

Sakamoto K, Kamimura M, Kurozumi S, Ito S (1995) Prostaglandin F2 alpha receptor. J Lipid Mediat Cell Signal 12:405–411

Sano K, Tanihara H, Heimark RL, Obata S, Davidson M, St John T, Taketani S, Suzuki S (1993) Protocadherins: a large family of cadherin-related molecules in central nervous system. EMBO J 12:249–2256

Schlaggar BL, Fox K, O'Leary DDM (1993) Postsynaptic control of plasticity in developing somatosensory cortex. Nature 364:623–626

Seeds NW, Williams BL, Bickford PC (1995) Tissue plasminogen activator induction in Purkinje neurons after cerebellar motor learning. Science 270:1992–1994

Sellmayer A, Uedelhoven WM, Weber PC, Bonventre JV (1991) Endogenous non-cyclooxygenase metabolites of arachidonic acid modulate growth and mRNA levels of immediate-early response genes in rat mesangial cells. J Biol Chem 266:3800–3807

Sharp FR, Sagar SM, Hicks K, Lowenstein D, Hisanaga K (1991) c-fos mRNA, Fos, and Fos-related antigen induction by hypertonic saline and stress. J Neurosci 11:2321–2331

Shatz CJ (1990) Impulse activity and the patterning of connections during CNS development. Neuron 5:745–756

Shatz CJ, Stryker MP (1988) Prenatal tetrodotoxin infusion blocks segregation of retinogeniculate afferents. Science 242:87–89

Sheng M, Greenberg M (1990) The regulation and function of c-fos and other immediate early genes in the nervous system. Neuron 4:477–485

Smith WL, Marnett LJ, DeWitt DL (1991) Prostaglandin and thromboxane biosynthesis. Pharmacol Ther 48:153–179

Stappert J, Kemler R (1994) A short core region of E-cadherin is essential for catenin binding and is highly phosphorylated. Cell Adhesion Commun 2:319–327

Steward O (1997) mRNA localization in neurons: a multipurpose mechanism? Neuron 18:9–12

Steward O, Wallace CS, Lyford GL, Worley PF (1998) Synaptic activation causes the mRNA for the IEG Arc to localize selectively near activated postsynaptic sites on dendrites. Neuron 21: 741–751

Tan A, Moratalla R, Lyford GL, Worley P, Graybiel AM (2000) The activity-regulated cytoskeletal-associated protein arc is expressed in different striosome-matrix patterns following exposure to amphetamine and cocaine. J Neurochem 74:2074–2078

Tretter YP, Hertel M, Munz B, ten Bruggencate G, Werner S, Alzheimer C (2000) Induction of activin A is essential for the neuroprotective action of basic fibroblast growth factor in vivo. Nat Med 6:812–815

Trudeau VL, Theodosis DT, Poulain DA (1997) Activin facilitates neuronal development in the rat amygdala. Neurosci Lett 237:33–36

Tsui C, Copeland NG, Gilbert DJ, Jenkins NA, Barnes CA, Worley PF (1996) Narp, a novel member of the pentraxin family, promotes neurite outgrowth and is dynamically regulated by neuronal activity. J Neurosci 16:2463–2478

Tsumoto T (1992) Long-term potentiation and long-term depression in the neocortex. Prog Neurobiol 39:209–228

Tsumoto T, Hagihara K, Sato K, Hata Y (1987) NMDA receptors in the visual cortex of young kittens are more effective than those of adult cats. Nature 327:513–514

Tu JC, Xiao B, Naisbitt S, Yuan JP, Petralia RS, Brakeman P, Doan A, Aakalu VK, Lanahan AA, Sheng M, Worley PF (1999) Coupling of mGluR/Homer and PSD-95 complexes by the Shank family of postsynaptic density proteins. Neuron 23:583–592

Uchida N, Honjo Y, Johnson KR, Wheelock MJ, Takeichi M (1996) The catenin/cadherin adhesion system is localized in synaptic junctions bordering transmitter release zones. J Cell Biol 135:767–779

Valentin E, Lambeau G (2000) Increasing molecular diversity of secreted phospholipases A(2) and their receptors and binding proteins. Biochim Biophys Acta 1488:59–70

Van den Pol AN (1994) Metabotropic glutamate receptor mGluR1 distribution and ultrastructural localization in hypothalamus. J Comp Neurol 349:615–632

van den Pol AN, Romano C, Ghosh P (1995) Metabotropic glutamate receptor mGluR5 subcellular distribution and developmental expression in hypothalamus. J Comp Neurol 362:134–150

Wallace CS, Lyford GL, Worley PF, Steward O (1998) Differential intracellular sorting of immediate early gene mRNAs depends on signals in the mRNA sequence. J Neurosci 18:26–35

Wang HW, Sui S (1999) Pentameric two-dimensional crystallization of rabbit C-reactive protein on lipid monolayers. J Struct Biol 127:283–286

Waters RS, McCandlish CA, Cooper NG (1990) Early development of SI cortical barrel subfield representation of forelimb in normal and deafferented neonatal rat as delineated by peroxidase conjugated lectin, peanut agglutinin (PNA). Exp Brain Res 81:234–240

Williams K, Russell SL, Shen YM, Molinoff PB (1993) Developmental switch in the expression of NMDA receptors occurs in vivo and in vitro. Neuron 10:267–278

Wong ROL, Meister M, Shatz CJ (1993) Transient period of correlated bursting activity during development of the mammalian retina. Neuron 11:923–938

Worley PF, Cole AJ, Saffen DW, Baraban JM (eds) (1990a) Transcription factor regulation in brain: focus on activity and NMDA dependent regulation. Springer, Berlin Heidelberg New York

Worley PF, Cole AJ, Murphy TM, Christy BA, Nakabeppu Y, Baraban JM (1990b) Synaptic regulation of immediate early genes in brain. Cold Springs Harb Symp Quant Biol 55:213–223

Worley PF, Christy BA, Nakabeppu Y, Bhat RV, Cole AJ, Baraban JM (1991) Constitutive expression of zif268 in neocortex is regulated by synaptic activity. Proc Natl Acad Sci USA 88: 5106–5110

Xiao B, Tu JC, Petralia RS, Yuan JP, Doan A, Breder CD, Ruggiero A, Lanahan AA, Wenthold RJ, Worley PF (1998) Homer regulates the association of group 1 metabotropic glutamate receptors with multivalent complexes of homer-related, synaptic proteins. Neuron 21:707–716

Xiao B, Tu JC, Worley PF (2000) Homer: a link between neural activity and glutamate receptor function. Curr Opin Neurobiol 10:370–374

Yamagata K, Andreasson K, Kaufmann WE, Barnes CA, Worley PF (1993) Expression of a mitogen-inducible cyclooxygenase in brain neurons: regulation by synaptic activity and glucocorticoids. Neuron 11:371–386

Yamagata K, Sanders LK, Kaufmann WE, Barnes CA, Nathans D, Worley PF (1994) *Rheb*, a growth factor and synaptic activity regulated gene, encodes a novel Ras-related protein. J Biol Chem 269:16333–16339

Yamagata K, Andreasson K, Sugiura H, Maru E, Irie Y, Miki N, Hayashi Y, Yoshioka M, Kaneko K, Kato H, Worley PF (1999) Arcadlin is an activity-regulated cadherin involved in long-term potentiation. J Biol Chem 274:19473–19479

Yee WM, Worley PF (1997) Rheb interacts with Raf-1 kinase and may function to integrate growth factor- and protein kinase A-dependent signals. Mol Cell Biol 17:921–933

Yermakova AV, Rollins J, Callaham LM, Rogers J, O'Banion MK (1999) Cyclooxygenase-1 in human Alzheimer and control brain: quantitative analysis of expression by microglia and CA3 hippocampal neurons. J Neuropathol Exp Neurol 58:1135–1146

Yuste R, Nelson DA, Rubin WW, Katz LC (1995) Neuronal domains in developing neocortex: mechanisms of coactivation. Neuron 14:7–17

Yuste R, Peinado A, Katz LC (1992) Neuronal domains in developing neocortex. Science 257: 665–669

Zahedi K (1996) Characterization of the binding of serum amyloid P to type IV collagen. J Biol Chem 271:14897–14902

Zhuo M, Holtzman DM, Li Y, Osaka H, DeMaro J, Jacquin M, Bu G (2000) Role of tissue plasminogen activator receptor LRP in hippocampal long-term potentiation. J Neurosci 20: 542–549

Role of Afferent Activity in the Development of Cortical Specification

ALVIN W. LYCKMAN[1] and MRIGANKA SUR[1,2]

Summary. The surgical cross-modal rewiring paradigm is an experimental method for examining the physiological and anatomical consequences of exposing developing cortical subregions to specific types of patterned sensory inputs. Data from these experiments provide strong inferences about the role of extrinsic (subcortical) cortical inputs in shaping the local cortical networks that organize and process sensory information. Behavioral results from this work also suggest that such activity (and activity in general) is a profound organizer of cerebral connectivity. We discuss one future direction of these studies: the implication that extrinsic inputs regulate developmental genes that are responsible for refining the connectivity within local circuits, and a strategy to discover and characterize such genes.

1 Introduction

Throughout the course of its development, casual inspection of the cortex suggests that it is rather uniform in cellular organization. Nonetheless, we know from neuroanatomy that different cortical areas receive very different types of inputs, and from neurophysiology that different regions of the cortex perform complex calculations that are highly specific to the nature of the inputs. More fine-grained anatomical and physiological studies strongly suggest that the complex responses of cortical neurons to input-specific processing depend on the precise patterning of local intracortical networks. How, then, during development, are these local networks patterned? To what extent can cortical regions develop local networks based solely on cues intrinsic and specific to local regions? Does cortex rely on extrinsic inputs to instruct the differentiation of local intracortical networks? In this chapter we will discuss the nature of mature cortical regionalization and then relate recent evidence that indicates that switching cortical inputs early in development can transform developing cortical networks so as to be able to appropriately process novel inputs. We speculate how extrinsic inputs might regulate the expression of cortical genes that control connectivity locally within intracortical networks.

[1] The Center for Learning and Memory, and
[2] the Department of Brain and Cognitive Sciences, Massachusetts Institute of Technology, E25-235, Cambridge, Massachusetts 02139, USA

Results and Problems in Cell Differentiation, Vol. 39
C.F. Hohmann (Ed.): Cortical Development

2 Sensory Modalities: Vision and Audition

Categories of information within the nervous system are broadly termed "modalities". Some obvious modalities are those derived from sensory epithelia that transduce particular external stimuli such as light, sound, contact, etc. These primary modalities are transmitted in pure form, but after some processing, to the specific *primary* cortical areas, where increasingly abstract types of processing by local intracortical circuits occur. Thereafter, the information within these separate modality streams diverge, with some pathways restricted to the original modality and some merging, taking on multimodal characteristics. Within these "higher" cortical areas, sensory information is further transformed and recombined, in ways that are less intuitive and highly abstract. It is likely that these higher areas are essential to our advanced cognitive and associative faculties. In order for these higher areas to function appropriately, information must be transferred in complex networks with the utmost order. We are only just beginning to describe some of these complex circuits and to appreciate how some of this order arises.

2.1 Visual Processing

Because of the intuitive organization of many of its cellular elements (Kuffler et al. 1984), and some experimental serendipity (Hubel and Wiesel 1998), the visual system is the sina qua non of functional neuroanatomy (Fig. 1A). Subcortically, the visual world is represented twice in a series of two-dimensional

Fig. 1A, B. Visual and auditory cortex. **A** Visual cortex. By the time visual information reaches V1, three two-dimensional maps of visual space have been formed. But, whereas neurons in the retina and the dLGN respond vigorously to points of contrast in space, neurons in V1 are exquisitely sensitive to edges at particular orientations. Functional mapping of intrinsic signals by optical imaging shows the organization of orientation selectivity within V1. As shown in the *oval*, the cortical surface can be divided into domains in which neighboring neurons respond best to edges of particular orientations – shown as *separately colored regions*. These domains cycle consistently and progressively at the "pinwheel centers", the points where all iso-orientation domains meet. Not shown, but also superimposed on the topographic map of visual space, as well as over the map of orientation selectivity, are maps of other anatomical features, such as ocular dominance, directional selectivity, and spatial frequency. **B** By the time auditory information reaches A1, it has passed through several tonotopically organized subcortical regions (not shown), the last of which is the *MGN*. A1 is also tonotopically organized: a given frequency is represented along parallel, one-dimensional isofrequency contours (*purple-indigo-cyan-green in the cartoons*). At least one cortical auditory receptive field property maps orthogonally to these contours: binaural response. In the *EE* bands, cortical neurons are excited by stimulus from both ears, while, in the alternating *EI* bands, cortical neurons are typically excited by the contralateral ear and inhibited by the ipsilateral ear. These cartoons illustrate some features of visual and auditory cortices that indicate very different underlying local cortical networks in the two systems

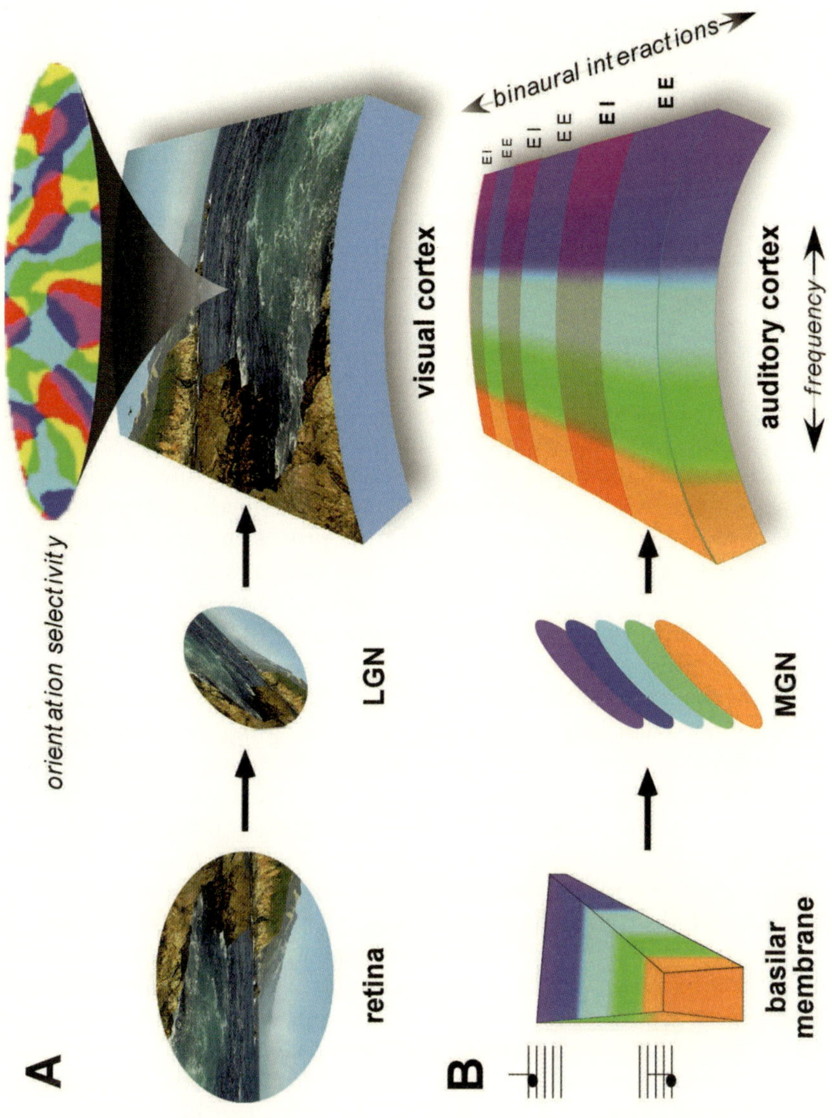
A
retina
LGN
visual cortex
orientation selectivity
B
basilar membrane
MGN
auditory cortex
frequency
binaural interactions
EI
EE
EI
EE
EI
EE

Fig. 1A, B

maps of light intensity and contrast. The first such map occurs in the retina, where a multilayered neuroepithelium transduces the light that enters the eye into a point-by-point map of contrast that resides in a monolayer of projection neurons, the retinal ganglion cells. The axons of the retinal ganglion cells project to the dorsal nucleus of the lateral geniculate of the thalamus (dLGN) where they form a map of retina, and hence the second map of visual space. The axons of the projection neurons of the dLGN project to primary visual cortex (V1), forming a third map of visual space. It is within this third map that processing of visual information generates new kinds of information that are not simply the presence or absence of contrast, but rather the orderly associations of congruous regions of space as oriented edges of contrast, edges of contrast borders, and directions of movement of edges of contrast. Thus, within V1, columns of cortical neurons respond best to oriented bars of light or dark, to such bars of specific lengths, or to such bars when moving in particular directions and speeds. Therefore, by the time visual information leaves V1, the dynamic, curvilinear visual world has been transformed into a code based on short line segments. Further processing leads to complex neural representations of objects and their surroundings, which cannot be dealt with in depth here – nonetheless, we will come back to these areas at the end of the chapter as our work may have ramifications for how connectivity might develop in these areas as well.

While visual processing in V1 was first described via classical methods of extracellular recording and receptive field mapping, a more complete view of the anatomical order underlying this cellular physiology was elegantly revealed using the technique of optical imaging of intrinsic cortical signals (Blasdel and Salama 1986; Grinvald et al. 1986; Frostig et al. 1990; Ts'o et al. 1990; Rao et al. 1997; Fig. 1A, inset). This technique generated maps that showed to a degree never previously appreciated that several classes of receptive field properties in V1 neurons generally changed in a progressive and orderly fashion across the whole of V1, and that these maps overlap (Hubener et al. 1997; Issa et al. 2000; White et al. 2001). The overlap of these maps allows the many different receptive field properties of V1 neurons to be equivalently represented for all areas of the central visual field (Swindale et al. 2000). What gives rise to the emergent receptive field characteristics of V1 neurons? Clearly, and as first described by Hubel and Wiesel (1998), an edge can be represented by a V1 neuron that receives inputs from a row of retinal ganglion cells. In that scenario, in order to represent all orientations at all points in visual space, dLGN projection neurons would be required to send seemingly myriad divergent projections to V1 neurons, with the constraint that retinotopic order be preserved. Moreover, an orderly progression of receptive field properties across V1 would seem to require that the patterning of the thalamocortical divergence show a repetitive characteristic across V1. There is experimental evidence that receptive field properties in V1 are in fact highly dependent on ordered thalamocortical projections (Chapman et al. 1991; Reid and Alonso 1995; Ferster et al. 1996). However, other experiments, including modeling data, suggest that activity in local intracortical networks helps to shape and sharpen the receptive field tuning of V1 neurons, and that perturbation of local intracortical networks can shift and/or weaken V1 receptive field maps (Crook and Eysel 1992; Allison et al. 1995; Somers et al. 1995; Crook et al. 1996; Dragoi and Sur 2000; Pugh et al. 2000). Thus, the emergence of orderly maps of V1 receptive field properties requires not only precision in the array of thalamocortical inputs, but also appropriate connectivity within complex local intracortical networks (Kozloski et al. 2001).

2.2 Auditory Processing

While perhaps less intuitive, the neural encoding of the aural world (Fig. 1B) is nonetheless as elegant and, at comparable anatomical stages, perhaps rather more profound than that found in the visual system (Willott 1983). Sound pressure levels of different frequencies maximally vibrate the basilar membrane of the cochlea at specific points. These vibrations are transduced by hair cells that are biased to respond to particular best frequencies and that in turn excite the axons of the spiral ganglion cells, whose axons form the auditory nerve. The ensemble activity in the auditory nerve therefore represents the activity along a linear dimension (representing frequency) of the basilar membrane, in essence, a Fourier transform of the sound pressure levels entering the ear. By the time this information reaches the subcortical auditory relay nucleus, the medial geniculate nucleus of the thalamus (MGN), it has been processed (in several subcortical relay stations) to encode several frequency-by-frequency emergent features of the original auditory stimulus, including interaural time difference and interaural intensity difference: features that are used to encode a map of auditory space. In addition, spectral integration, wherein subsets of auditory neurons respond best to combinations of pure tones, occurs at subcortical levels, beginning the task of identifying an auditory stimulus.

Optical imaging of intrinsic signals in the primary auditory cortex (A1) confirms that the pattern of innervation of A1 by relay neurons of the MGN continues to respect the tonotopic organization, i.e., mapping of the basilar membrane, found in all the subcortical stations of the auditory system (Bakin et al. 1996; Dinse et al. 1997). Standard extracellular electrophysiology has demonstrated some degree of topographic regularity in other receptive field properties for A1 neurons. For instance, binaural interactions (coexcitation by the two ears or antagonistic excitation-inhibition by the two ears) map orthogonally to isofrequency contours (Middlebrooks et al. 1980), while neurons subserving spectral integration (response modulation by simultaneous presentation of non-best frequencies) are organized into periodic patches (Read et al. 2001). Although it is still unclear how many other auditory receptive field properties are topographically organized in A1 (Mendelson and Grasse 1992; Kilgard and Merzenich 1999; Brosch and Schreiner 2000; Schreiner et al. 2000), it is likely there will be an underlying topographic organization for them as well. What the existing data on topographic organization of receptive field properties make clear is that there is virtually no resemblance between V1 and A1.

2.3 Vision Versus Audition

In comparing some basic functions of two sensory systems, vision and audition, such as localization of stimulus with respect to the organism, and its identification based on characteristic features, we see that the two systems must operate quite differently. For stimulus localization, vision can rely simply on activity in retinotopic maps and, complications of form and feature binding aside, it is quite intuitive that movement of a stimulus will simply activate a contiguous field of activity within the retinotopic maps. For audition, stimulus localization requires calculating frequency-specific interaural features along with some frequency spectral analyses, i.e., sets of calculations that, although comprehensible, are at least a step removed from the point-to-point representation afforded by the visual system. Identification of either visual or auditory stimuli is certainly not trivial, and in both modalities we see that, at the earliest stages, the neural strategy involves feature abstraction: visual elements become sets of edges, and auditory elements become sets of frequency combinations. However, the precise nature of representing light versus sound pressure level involves completely different neural processing mechanisms. We surmise that sharpening and orderly arrangements of multifarious auditory receptive fields is as dependent on local intracortical networks as appears to be the case for the different receptive field classes of V1 neurons. And thus, it seems unlikely that the local intracortical networks that are peculiar to V1 and to A1 could substitute for each other (see Fig. 1). Rather, although we are currently far from having complete and accurate descriptions of these connections, we can confidently surmise that the local intracortical networks in V1 and A1 are quite different. How do these different circuits arise?

3 Intrinsic Determination of Modality-Specific Subregions of Cortex

Visual cortex (V1) is situated in posterior, temporal cortex, while A1 sits laterally, encircled by the ectosylvian sulcus. Histological analyses of developing embryonic and neonatal cortex have revealed the presence of several cortex-wide molecular gradients that are generated by locally regulated transcription of developmental genes (Rubenstein et al. 1999). There is currently, however, no evidence that any one gene serves to exclusively specify the connectivity of any cortical subregion. In the absence of modality-specific labels, the identification of several molecular gradients has led to the suggestion that these gradients might interact to specify unique cortical subregions and attract modality-specific sets of thalamic inputs (Bishop et al. 2000). Studies of mice transgenic for at least two such gradients show shifted expression of other cortical gradients and show altered thalamocortical connectivity in particular cortical subregions (Mallamaci et al. 2001). Embryonic transplantation experiments, wherein transplanted cortical regions retain the ability to form cortical connections characteristic of their original location, confirm an early heterogeneity of the cortical mantle that is intrinsically determined (Gaillard and Roger 2000; Pinaudeau et al. 2000). However, is this early intrinsic specification sufficient to determine how local intracortical circuits are laid out? And, can intrinsic specification be over-ridden by extrinsic inputs carrying information in modalities not normally presented to a given cortical subregion?

4
A Role for Extrinsic Inputs in Specification of Local Cortical Networks

4.1
Theoretical Considerations for Experimentally Altering Cortical Inputs

As discussed above, mature cortical circuits, replete with local, recurrent excitatory and inhibitory connections between neurons that are downstream of thalamic inputs, and confer integrity and order to receptive field organization in cortical subregions. In order to determine whether activity within particular thalamocortical projections is responsible for shaping these modality-specific local networks, one might ask whether such networks develop in the *absence* of activity, perhaps as would occur if a particular thalamic relay, say the dLGN, were eliminated early in development. However, activity in these local circuits will also be driven by inputs not only from subcortical relays, but also between different cortical subregions, via so-called long-range connections. Thus, there is no theoretical exigency for thalamic inputs to be the sole sculptors of modality-specific local circuits. Moreover, since the development of these circuits, and of many others as well, depends on activity per se (Shatz and Stryker 1988; Sretavan et al. 1988; Catalano and Shatz 1998), finding changes in these networks after removing activity does not inform as to whether the patterning of local networks depends specifically on the nature of the input, i.e., its dynamic *spatiotemporal structure.*

An instructional, rather than merely permissive, role for extrinsic inputs can only be assessed by *changing* them, rather than removing them. Experimentally, this is best achieved by exchanging one input source with another, because one can, for a given cortical subregion with altered inputs, compare its local network *not only* to the normal local network, *but also* to the local network normally associated with the new input. In such experiments, there are two theoretical considerations related to input switching. First, and more trivially, how quantitatively complete is the switching: is there any contamination from the original source, and is the new source capable of completely replacing the original? Second, and fundamentally, how different are the original source and the new source in terms of the spatiotemporal patterning characteristics? Are the patterns of the different inputs really peculiar to their respective sources and distinct from one another? At the simplest level of analysis, if efficient conversion of input sources can be assumed, and the two sources generate genuinely distinct activity patterns, then the properties of the local network that are stable and obtainable with both sources are likely to derive from developmental mechanisms inherent in the cortical subregion, hence input-independent. However, any new properties, developed by a cortical region that receives a novel input, that are *shared with the local network that normally receives that same input* are likely to derive from a common developmental response to some feature(s) peculiar to that input source.

To date, experiments that have employed the *cross-modal rewiring paradigm* are the closest we have come to meeting these theoretical requirements. We will discuss data from these experiments to make the case that there is compelling evidence that local cortical networks are profoundly respecified by features characteristic to their extrinsic inputs.

4.2
The Rewiring Paradigm

In the surgical rewiring paradigm (Fig. 2), first used by Gerald Schneider at the Massachusetts Institute of Technology, retinal ganglion cell axons, which normally innervate the dLGN, are induced to innervate the medial geniculate body (MGB) postnatally, but still early with respect to cortical development (Schneider 1973; Frost 1981; Sur et al. 1988). In the original paradigm, a major subcortical target of retinal ganglion cell axons, the superior colliculus, is eliminated. In addition, the normal source of innervation to the MGB, the inferior colliculus, is also eliminated. This leaves retinal ganglion cell axons without normal target space, but also immediately adjacent to a denervated, albeit novel, target field. Recently, we have shown that a cue(s) emitted from the denervated MGB is, in fact, sufficient to induce retinal ganglion cell axons to innervate the denervated zone (Lyckman et al. 2001). In animals having undergone this procedure, synapses form between retinal ganglion cell axons and neurons of the MGB that transmit visual information into the ascending auditory stream. It is crucial to recognize that the surgery leaves the ascending thalamocortical pathway between the MGB and A1, with its tonotopic mapping, patent and untouched – only now, it transmits visually driven activity along the same "wires" that would normally carry auditory traffic (see Fig. 2B). As a secondary issue, one might ask why retinal ganglion cells do not normally innervate the MGB? We think the answer lies partly in the fact that barriers exist that specifically prohibit retinal ganglion cell axons from innervating MGB neurons (Fig. 2C,D), and we have recently shown that removing some of these barriers enhances the rewiring effect (Lyckman et al. 2001).

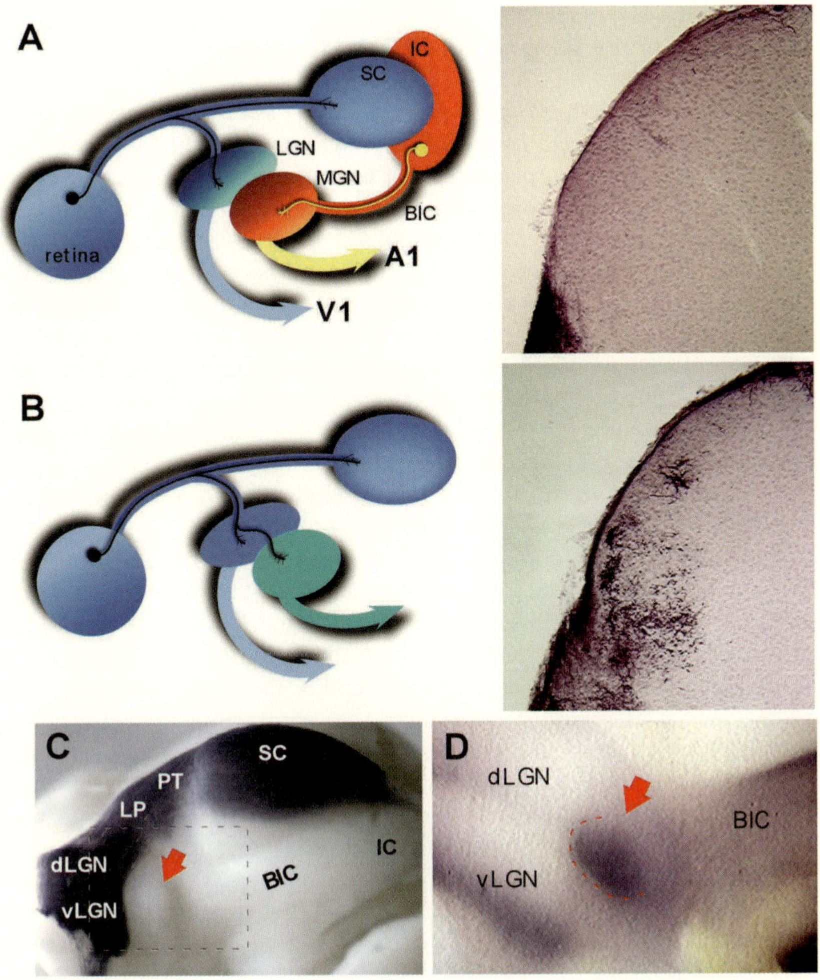

Fig. 2A–D. Rewiring of mammalian brain. **A** Normal visual and auditory pathways. *Left* The retina sends ascending visual information primarily to the dLGN and the superior colliculus (*SC*). The *MGN* receives ascending auditory information from the inferior colliculus (*IC*). *Right* Coronal section of the mouse MGN, showing the absence of retinal innervation in the normal MGN. **B** Rewiring of visual and auditory pathways. *Left* In neonatal rewiring surgery, ablation of the IC is sufficient to induce the retina to innervate the MGN – presumably because of a signal released by the denervated MGN that is attractive to growing retinal ganglion cell axons. In most published studies, the SC was also ablated to reduce the available target territory available to growing retinal ganglion cell axons. *Right* Coronal section of the MGN in a rewired mouse, innervated by retinal ganglion cell axons (*dark staining*). **C** Lateral view (posterior to the right) of the normal neonatal mouse optic tract, showing retinal ganglion cell axons (*black staining*) passing along the border of the *LGN* and the *MGN* (*arrow*). **D** Lateral view corresponding to the boxed region in **C** of the thalamus stained for ephrin-A proteins. These proteins form a repulsive barrier at the border between the LGN and the MGN – in the absence of this molecular barrier, rewiring in the mouse is increased (see Lyckman et al. 2001 for details)

4.3
Innervation of the Denervated MGN by the Retina

In order to understand how the rewiring paradigm specifically addresses various factors involved in specification of local cortical networks by extrinsic inputs, it is necessary to appreciate, first of all, that the rewired pathway is relatively sparse – in rewired ferrets, retinal ganglion cell axons innervate varying fractions of the MGN (Angelucci et al. 1998; Lyckman et al. 2001). In mice, we have shown that rewiring can be enhanced approximately three-fold in certain transgenic strains as compared to wild-type strains (Lyckman et al. 2001). A large amount of MGN is not innervated by either auditory or visual inputs, and we have not fully investigated what other inputs, if any, might also succeed in innervating the MGN after rewiring. Second, in ferrets, the terminal arbors of individual retinal ganglion cells, rather than retaining their isotropic distributions typically found in the dLGN, show elongation along the isofrequency axis in MGN, as if retinal axons can respond to guidance cues within MGN (Pallas et al. 1994; Pallas and Sur 1994).

Third, both eyes contribute to the rewired projection, and, in ferrets (Angelucci et al. 1997) and mice (Ellsworth et al. 2001), the inputs from the two eyes, remarkably, remain segregated. While we do not know the mechanism of this segregation, in rewired ferrets, retinal ganglion cell arbors in MGB are initially rather diffuse and later become more focused and condensed (Angelucci et al. 1997). It is possible that some time- or activity-dependent processes may contribute to eye-specific segregation. Finally, retinal ganglion cell arbors in ferrets show crude two-dimensional retinotopic order in the MGN (Roe et al. 1991). We have recently shown that the same molecular family, the ephrins, that is expressed in gradients in LGN and that contributes to mapping of at least one axis of the retino-LGN map, is also expressed as similar gradients in MGN (Lyckman et al. 2001) – so that retinotopy in the retino-MGN projection and in the retino-LGN projection is probably due to some common mechanisms. Thus, the rewired retino-MGN projection has not only characteristics of normal auditory inputs (isofrequency orientation of terminal arbors), but also characteristics of normal retino-LGN projections (eye-specific segregation and retinotopy).

4.4
Physiological Consequences of Rewiring

Rewiring surgery profoundly alters physiological inputs to A1 but not their anatomical organization. In the case of bilateral rewiring surgery, there is absolutely no auditory input to cortex, whereas unilateral rewiring surgery results in complete elimination of ascending auditory input to A1 only on the operated side of the brain. In the unilateral cases, however, intact callosal connections would permit A1 on the rewired side to receive auditory corticocortical inputs from A1 on the unlesioned side of the brain. It will be interesting to discover how these corticocortical connections regulate the development of local networks in A1. Within days of rewiring surgery, the lost auditory traffic to A1 is replaced by a substantial and effective (see below) amount of visual traffic along the normal thalamocortical (MGN→A1) pathway. Recall, again, that the anatomical patterning of MGN-A1 connections, whether normal or rewired, is *tonotopic*, hence very different from the *retinotopic* patterning of DLGN-V1 connections. In essence, what changes in the rewiring paradigm is the transformation of the *temporal structure* of the ascending sensory traffic to A1, from an auditory pattern to a visual pattern. Does this effect cortical development? Does rewired A1 resemble normal A1 or normal V1?

4.5 Analyses of Rewired A1

Three basic techniques have been employed to functionally and anatomically characterize neuronal connectivity in the rewired auditory cortex of the ferret. These data have been extensively reviewed (Sur et al. 1999; Sharma et al. 2000; Sur and Leamey 2001) and only the salient features germane to this discussion will be covered here.

4.5.1 Receptive Field Mapping

First, classical receptive field mapping by extracellular recording has shown that large numbers of neurons in rewired A1 are strongly driven by visual stimulation much as occurs for neurons in V1, i.e., rewired A1 neurons are in retinotopic maps, and often show strong preference for orientation and direction of visual stimuli (Sur et al. 1988; Roe et al. 1991, 1992; Sharma et al. 2000). Whereas neurons in V1 are located in vertical columns that share the same topographic representation and orientation selectivity, the same is true for a large percentage of neurons in rewired A1. Sharing of receptive field properties by neurons within a given column of cortex suggests that they share common inputs and/or form part of highly organized local networks that confer similar receptive field properties to sets of neighboring neurons (Allison et al. 1995; Mountcastle 1998). Orientation selectivity arises from convergence or inputs from rows of retinal ganglion cells. Thus, despite the tonotopic organization of the MGN-A1 projection, the MGN inputs to rewired A1 neurons are organized much as dLGN inputs converge onto V1 neurons. This could occur if the retinotopy of the retino-MGN projection is at least partially preserved in the rewired MGN-A1 projection, in conjunction with visually driven activity-dependent rearrangements of MGN-A1 connections. It will be interesting to explore whether thalamocortical arbors undergo such rearrangements during development of the rewired projection.

4.5.2 Optical Imaging of Intrinsic Signals

Optical imaging of intrinsic signals was used to examine the overall organization of orientation selectivity of visually driven neurons in rewired A1 of ferrets (Sharma et al. 2000). Again, organization of receptive field properties strongly indicates the presence of an orderly, local network that reinforces receptive field tuning and organizes neurons into local pools with similar receptive field characteristics. Optical imaging data show clearly that visually responsive neurons in rewired A1 are organized according to their orientation tuning, with nearest neighbors sharing receptive field properties (Sharma et al. 2000). Thus in V1, as in rewired A1, neurons with similar orientation tuning are clustered in domains, and the tuning in adjacent domains changes progressively as they circle around a so-called pinwheel center. That this pattern, the hallmark of organization of orientation selectivity in V1 (Rao et al. 1997), is repeated in rewired A1 indicates the profound influence of extrinsic inputs in modeling the circuitry of a local cortical network. These data leave the strongest impression that developing cortex is highly dependent on external instruction in the specification of its internal connectivity.

The optical imaging data also suggest that there are limits to the extent to which extrinsic signals can mold local cortical networks. First of all, the iso-orientation domains are larger in rewired A1 than in V1, suggesting less dense representation of a wide range of orientations. This is perhaps accounted for by the relative sparseness of visual information in the rewired MGN-A1 projection compared to the abundance of that in the dLGN-V1 projection. In addition, there is reduced regularity in the patterning of iso-orientation domains in rewired A1 as compared to V1. This might suggest that the local network in A1 is somewhat less capable of responding to the temporal structure of visual traffic in consolidating a repeating anatomical pattern of connections. Reduced regularity of the orientation map in the face of the presence of sharp orientation tuning suggests that these two features of V1 could be mechanistically dissociated.

4.5.3 Local Connections Within A1

Bulk injections of neuronal tracers into the cortical parenchyma can be used to reveal patterns of connectivity across relatively small domains of cortical subregions. For instance, biocytin microinjected into an iso-orientation domain of a given orientation in normal V1 labels cell bodies and axonal terminals that are uniformly (isotropically) distributed around the site of injection (Boyd and Matsubara 1991; Kisvarday and Eysel 1992). In addition, labeling of cell bodies and axonal terminals is also seen in nearby iso-orientation domains of the *same* orientation (Gilbert and Wiesel 1989; Kisvarday and Eysel 1992). This shows that the underlying local network is dominated by connections within and between same-orientation domains. In normal A1, however, experiments show that corticocortical connections show labeling of cell bodies and axonal terminals along linearly oriented isofrequency bands (Matsubara and Phillips 1988; Luethke et al. 1989; Ojima et al. 1991) – a very different pattern of labeling, revealing a very different underlying cortical network in A1 versus V1. What kind of network is revealed in rewired A1 using the same technique? Quantitative measures of HRP-labeling patterns in rewired A1 are intermediate between those of normal A1 and normal V1, suggesting that rewired local networks are a hybrid of the other two forms (Sharma et al. 2000). These anatomical data strongly indicate that the effects of switching inputs result not only in physiological and functional alterations of cortical subregions, but also that altering inputs changes the growth characteristics of the cellular elements of the network.

4.6 Other Signaling Mechanisms

Although we have a strong bias for thinking that electrical activity is the main player in altering local cortical networks in the rewiring paradigm, there is also the theoretical possibility that signaling molecules could be passed across synapses, and that such signaling molecules could independently regulate the development of cortical networks. Whether such signaling systems exist is unknown, but chemical transynaptic signaling has been implicated in other contexts, e.g., long-term potentiation (Fitzsimonds and Poo 1998; Tao et al. 2000; Kohara et al. 2001). In the case of the rewiring paradigm, a chemical signal could be passed from retinal ganglion cell axons to the relay neurons of the MGN, and in turn from their axons to cortical neurons – influencing their differentiation. In fact, transneuronal transport (from retina to cortex) of exogenous proteins such as lectin-coupled horseradish peroxidase, or of retinal proteins metabolically labeled by ^{3}H-proline, has been used to reveal the presence of eye-specific segregation of dLGN axons in V1 (LeVay and Voigt 1990; Antonini et al. 1999; Ruthazer et al. 1999). It is possible that molecules capable of specific developmental signaling, such as the neurotrophins, can be distributed to secondary targets via transneuronal transfer.

4.7 Behavior and Effects Downstream of Primary Sensory Cortical Areas

Behavioral data suggest that rewired cortices contribute to sensory function. Rewired ferrets are capable of using the rewired auditory cortex to respond to light as a visual, rather than auditory, stimulus (von Melchner et al. 2000). In addition, it has been shown that rewired hamsters can use rewired auditory cortex to discriminate visual stimuli (Frost et al. 2000). These findings strongly suggest that not only does rewiring alter local cortical networks in A1, but also that its outputs, and hence behavioral impact, become reprogrammed and associated with a new modality. In essence, the behavioral data suggest that the effects of neonatal rewiring cascade throughout the whole of the developing cerebrum, allowing new functional associations among disparate brain regions. If so, this inference indicates that activity is a profound organizer during development; that the plethora of medium, long-range and hemispheric connections are rather unpredetermined as to functional significance; and, that, in creating the cortex, evolution has created an intensely adaptive and plastic developmental substrate.

4.8 Strategy to Identify and Characterize Cortical Genes Activated by Modality-Specific Inputs

We hypothesize that at the heart of the ability of visual inputs to induce A1 neurons to develop local circuits that process visual information are changes in the growth characteristics of A1 neurons – characteristics that control the shapes and extents of dendritic arbors, local axonal targeting, and placement of synapses, and that are subject to regulatory control by cellular signal transduction mechanisms coupled to the electrical activation of neurons. Simply put, visual inputs are capable of inducing cortical genes in ways that lead to the formation of V1-like circuits. Activity influences the expression patterns of many neuronal genes (Corriveau 1999; Nedivi 1999). In the case of activity-dependent expression of genes that are responsible for shaping cortical networks, the main question is: can supplying a given input to a new cortical area convert its gene expression pattern to the one associated with the cortical area that is the normal target of that input? In the case of the rewiring paradigm, we can specifically ask: (1) does the pattern of gene expression differ between normal A1 and rewired A1? and (2) is the pattern of gene expression in rewired A1 more similar to that of normal A1 or normal V1? Recent advances in high-throughput gene expression assays make asking such questions feasible.

DNA-microarray technology permits assessment of the relative strength of expression of tens of thousands of genes, virtually the entire genome of some species, with small tissue samples (Fig. 3). Our recent breakthrough has been to show that the rewiring paradigm induces retinal innervation of the MGN in the mouse (Lyckman et al. 2001; see Fig. 2), a species for which extremely comprehensive, commercially available DNA microarrays exist. Further, we have new data that support the conclusion that retinal innervation of MGN in mouse leads to visual activation of rewired A1 (Fig. 3A). We are now poised to determine how alteration of physiological inputs to a cortical subregion alters its pattern of gene expression. Further down that road, we will have to functionally characterize genes with altered expression patterns, by examining gene product expression patterns, and by gene targeting experiments using cells, tissues and animals. We expect that such studies will lead not only to a detailed schematic of genetic regulation of intracortical connectivity, but also to insights into organizing principles of developing neural networks. There is hope that

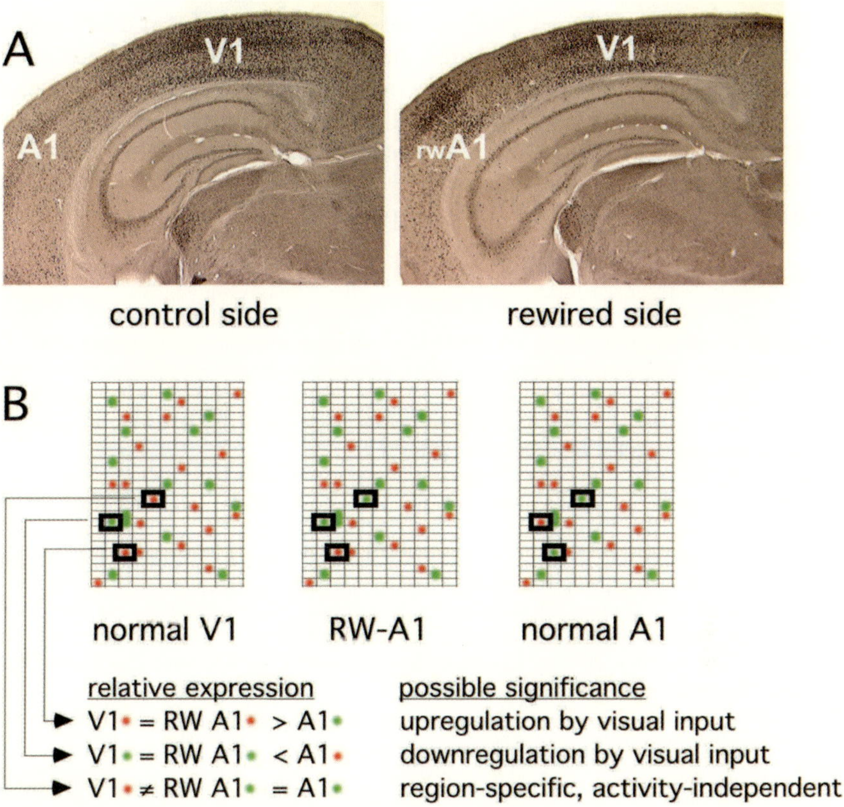

Fig. 3A, B. Experimental approach to discovering genes regulated specifically by patterned electrical activity in vivo. **A** Micrographs of c-fos immunohistochemistry in a mature rewired mouse: c-fos is an immediate early gene that is induced and maintained by synaptic activity. Here, we have light- and sound-deprived a rewired mouse for 3 days, to allow cortical expression of c-fos to diminish. Two hours prior to sacrifice, the mouse was given normal visual stimulation in its cage. *Left* The micrograph from the control (unoperated) side of the brain shows marked nuclear immunostaining of c-fos protein in neurons of V1 in response to light stimulation. A1, however, shows very low, background levels of staining. *Right* The rewired side of the brain shows strong c-fos immunostaining in V1, *but also in A1*, demonstrating the presence of functional visual connections to the rewired A1. **B** DNA-microarrays allow the quantitative assessment of gene expression for vast numbers of individual genes. Using arrays available for mouse, we have analyzed and compared the gene expression patterns found in normal V1, normal A1 and rewired A1. DNA-microarray analysis identifies those genes having relatively high (*red cells*) and low (*green cells*) level expression from those with average expression (*white cells*). From such data, it is possible to then identify genes whose level of expression differs significantly across different samples. We are particularly interested in genes which are highly (or lowly) expressed in V1 and rewired A1 as compared to normal A1 – such genes are candidates for regulation by visually patterned activity. Of course, the same technique can be used to identify cortical genes whose patterns of expression differ because of region-specific factors. Candidate genes need then be confirmed by expression and gene targeting experiments

information will also lead to therapies for congenital neural abnormalities related to sensory abnormalities and deficits as in Usher syndrome (Eudy and Sumegi 1999; Keats and Corey 1999). Finally, mature cortical areas are also able to respond to novel, cross-modal inputs when their normal inputs are removed or atrophy (Giraud et al. 2001; Levanen and Hamdorf 2001) – whether the ability of mature cortex to functionally process novel inputs is related to the regulation of gene expression by its extrinsic inputs awaits investigation.

References

Allison JD, Casagrande VA, Bonds AB (1995) The influence of input from the lower cortical layers on the orientation tuning of upper layer V1 cells in a primate. Vis Neurosci 12:309–320

Angelucci A, Clasca F, Bricolo E, Cramer KS, Sur M (1997) Experimentally induced retinal projections to the ferret auditory thalamus: development of clustered eye-specific patterns in a novel target. J Neurosci 17:2040–2055

Angelucci A, Clasca F, Sur M (1998) Brainstem inputs to the ferret medial geniculate nucleus and the effect of early deafferentation on novel retinal projections to the auditory thalamus. J Comp Neurol 400:417–439

Antonini A, Fagiolini M, Stryker MP (1999) Anatomical correlates of functional plasticity in mouse visual cortex. J Neurosci 19:4388–4406

Bakin JS, Kwon MC, Masino SA, Weinberger NM, Frostig RD (1996) Suprathreshold auditory cortex activation visualized by intrinsic signal optical imaging. Cereb Cortex 6:120–130

Bishop KM, Goudreau G, O'Leary DD (2000) Regulation of area identity in the mammalian neocortex by Emx2 and Pax6. Science 288:344–349

Blasdel GG, Salama G (1986) Voltage-sensitive dyes reveal a modular organization in monkey striate cortex. Nature 321:579–585

Boyd J, Matsubara J (1991) Intrinsic connections in cat visual cortex: a combined anterograde and retrograde tracing study. Brain Res 560:207–215

Brosch M, Schreiner CE (2000) Sequence sensitivity of neurons in cat primary auditory cortex. Cereb Cortex 10:1155–1167

Catalano SM, Shatz CJ (1998) Activity-dependent cortical target selection by thalamic axons. Science 281:559–562

Chapman B, Zahs KR, Stryker MP (1991) Relation of cortical cell orientation selectivity to alignment of receptive fields of the geniculocortical afferents that arborize within a single orientation column in ferret visual cortex. J Neurosci 11:1347–1358

Corriveau RA (1999) Electrical activity and gene expression in the development of vertebrate neural circuits. J Neurobiol 41:148–157

Crook JM, Eysel UT (1992) GABA-induced inactivation of functionally characterized sites in cat visual cortex (area 18): effects on orientation tuning. J Neurosci 12:1816–1825

Crook JM, Kisvarday ZF, Eysel UT (1996) GABA-induced inactivation of functionally characterized sites in cat visual cortex (area 18): effects on direction selectivity. J Neurophysiol 75: 2071–2088

Dinse HR, Godde B, Hilger T, Reuter G, Cords SM, Lenarz T, von Seelen W (1997) Optical imaging of cat auditory cortex cochleotopic selectivity evoked by acute electrical stimulation of a multi-channel cochlear implant. Eur J Neurosci 9:113–119

Dragoi V, Sur M (2000) Dynamic properties of recurrent inhibition in primary visual cortex: contrast and orientation dependence of contextual effects. J Neurophysiol 83:1019–1030

Ellsworth CA, Lyckman AW, Sur M (2001) Eye-specific patterning of retino-geniculate terminations in the medial geniculate nucleus of rewired mice. Soc Neurosci Abstr 27, No 14

Eudy JD, Sumegi J (1999) Molecular genetics of Usher syndrome. Cell Mol Life Sci 56:258–267

Ferster D, Chung S, Wheat H (1996) Orientation selectivity of thalamic input to simple cells of cat visual cortex. Nature 380:249–252

Fitzsimonds RM, Poo M-M (1998) Retrograde signaling in the development and modification of synapses. Physiol Rev 78:143–170
Frost DO (1981) Orderly anomalous retinal projections to the medial geniculate, ventrobasal, and lateral posterior nuclei of the hamster. J Comp Neurol 203:227–256
Frost DO, Boire D, Gingras G, Ptito M (2000) Surgically created neural pathways mediate visual pattern discrimination. Proc Natl Acad Sci USA 97:11068–11073
Frostig RD, Lieke EE, Ts'o DY, Grinvald A (1990) Cortical functional architecture and local coupling between neuronal activity and the microcirculation revealed by in vivo high-resolution optical imaging of intrinsic signals. Proc Natl Acad Sci USA 87:6082–6086
Gilbert CD, Wiesel TN (1989) Columnar specificity of intrinsic horizontal and corticocortical connections in cat visual cortex. J Neurosci 9:2432–2442
Giraud A, Price CJ, Graham JM, Truy E, Frackowiak RS (2001) Cross-modal plasticity underpins language recovery after cochlear implantation. Neuron 30:657–663
Grinvald A, Lieke E, Frostig RD, Gilbert CD, Wiesel TN (1986) Functional architecture of cortex revealed by optical imaging of intrinsic signals. Nature 324:361–364
Hubel DH, Wiesel TN (1998) Early exploration of the visual cortex. Neuron 20:401–412
Hubener M, Shoham D, Grinvald A, Bonhoeffer T (1997) Spatial relationships among three columnar systems in cat area 17. J Neurosci 17:9270–9284
Issa NP, Trepel C, Stryker MP (2000) Spatial frequency maps in cat visual cortex. J Neurosci 20: 8504–8514
Keats BJ, Corey DP (1999) The Usher syndromes. Am J Med Genet 89:158–166
Kilgard MP, Merzenich MM (1999) Distributed representation of spectral and temporal information in rat primary auditory cortex. Hear Res 134:16–28
Kisvarday ZF, Eysel UT (1992) Cellular organization of reciprocal patchy networks in layer III of cat visual cortex (area 17). Neuroscience 46:275–286
Kohara K, Kitamura A, Morishima M, Tsumoto T (2001) Activity-dependent transfer of brain-derived neurotrophic factor to postsynaptic neurons. Science 291:2419–2423
Kozloski J, Hamzei-Sichani F, Yuste R (2001) Stereotyped position of local synaptic targets in neocortex. Science 293:868–872
Kuffler SW, Martin AR, Nicholls JG (1984) From neuron to brain: a cellular approach to the function of the nervous system. Sinauer Associates, Sunderland, MA
Levanen S, Hamdorf D (2001) Feeling vibrations: enhanced tactile sensitivity in congenitally deaf humans. Neurosci Lett 301:75–77
LeVay S, Voigt T (1990) Retrograde transneuronal transport of wheat-germ agglutinin to the retina from visual cortex in the cat. Exp Brain Res 82:77–81
Luethke LE, Krubitzer LA, Kaas JH (1989) Connections of primary auditory cortex in the New World monkey, Saguinus. J Comp Neurol 285:487–513
Lyckman AW, Jhaveri S, Feldheim DA, Vanderhaeghen P, Flanagan JG, Sur M (2001) Enhanced plasticity of retinothalamic projections in an ephrin-A2/A5 double mutant. J Neurosci 21: 7684–7690
Mallamaci A, Muzio L, Chan CH, Parnavelas J, Boncinelli E (2000) Area identity shifts in the early cerebral cortex of Emx2–/– mutant mice. Nat Neurosci 3:679–686
Matsubara JA, Phillips DP (1988) Intracortical connections and their physiological correlates in the primary auditory cortex (AI) of the cat. J Comp Neurol 268:38–48
Mendelson JR, Grasse KL (1992) A comparison of monaural and binaural responses to frequency modulated (FM) sweeps in cat primary auditory cortex. Exp Brain Res 91:435–454
Middlebrooks JC, Dykes RW, Merzenich MM (1980) Binaural response-specific bands in primary auditory cortex (AI) of the cat: topographical organization orthogonal to isofrequency contours. Brain Res 181:31–48
Mountcastle VB (1998) Perceptual neuroscience: the cerebral cortex. Harvard University Press, Cambridge, MA
Nedivi E (1999) Molecular analysis of developmental plasticity in neocortex. J Neurobiol 41: 135–147

Ojima H, Honda CN, Jones EG (1991) Patterns of axon collateralization of identified supragranular pyramidal neurons in the cat auditory cortex. Cereb Cortex 1:80–94
Pallas SL, Sur M (1994) Morphology of retinal axon arbors induced to arborize in a novel target, the medial geniculate nucleus. II. Comparison with axons from the inferior colliculus. J Comp Neurol 349:363–376
Pallas SL, Hahm J, Sur M (1994) Morphology of retinal axons induced to arborize in a novel target, the medial geniculate nucleus. I. Comparison with arbors in normal targets. J Comp Neurol 349:343–362
Pinaudeau C, Gaillard A, Roger M (2000) Stage of specification of the spinal cord and tectal projections from cortical grafts. Eur J Neurosci 12:2486–2496
Pugh MC, Ringach DL, Shapley R, Shelley MJ (2000) Computational modeling of orientation tuning dynamics in monkey primary visual cortex. J Comput Neurosci 8:143–159
Rao SC, Toth LJ, Sur M (1997) Optically imaged maps of orientation preference in primary visual cortex of cats and ferrets. J Comp Neurol 387:358–370
Read HL, Winer JA, Schreiner CE (2001) Modular organization of intrinsic connections associated with spectral tuning in cat auditory cortex. Proc Natl Acad Sci USA 98:8042–8047
Reid RC, Alonso JM (1995) Specificity of monosynaptic connections from thalamus to visual cortex. Nature 378:281–284
Roe AW, Hahm J, Sur M (1991) Experimentally induced establishment of visual topography in auditory thalamus. Soc Neurosci Abstr 17:898
Roe AW, Pallas SL, Kwon YH, Sur M (1992) Visual projections routed to the auditory pathway in ferrets: receptive fields of visual neurons in primary auditory cortex. J Neurosci 12:3651–3664
Rubenstein JL, Anderson S, Shi L, Miyashita-Lin E, Bulfone A, Hevner R (1999) Genetic control of cortical regionalization and connectivity. Cereb Cortex 9:524–532
Ruthazer ES, Baker GE, Stryker MP (1999) Development and organization of ocular dominance bands in primary visual cortex of the sable ferret. J Comp Neurol 407:151–165
Schneider GE (1973) Early lesions of superior colliculus: factors affecting the formation of abnormal retinal projections. Brain Behav Evol 8:73–109
Schreiner CE, Read HL, Sutter ML (2000) Modular organization of frequency integration in primary auditory cortex. Annu Rev Neurosci 23:501–529
Sharma J, Angelucci A, Sur M (2000) Induction of visual orientation modules in auditory cortex. Nature 404:841–847
Shatz CJ, Stryker MP (1988) Prenatal tetrodotoxin infusion blocks segregation of retinogeniculate afferents. Science 242:87–89
Somers DC, Nelson SB, Sur M (1995) An emergent model of orientation selectivity in cat visual cortical simple cells. J Neurosci 15:5448–5465
Sretavan DW, Shatz CJ, Stryker MP (1988) Modification of retinal ganglion cell axon morphology by prenatal infusion of tetrodotoxin. Nature 336:468–471
Sur M, Leamey CA (2001) Development and plasticity of cortical areas and networks. Nat Rev Neurosci 2:251–262
Sur M, Garraghty PE, Roe AW (1988) Experimentally induced visual projections into auditory thalamus and cortex. Science 242:1437–1441
Sur M, Angelucci A, Sharma J (1999) Rewiring cortex: the role of patterned activity in development and plasticity of neocortical circuits. J Neurobiol 41:33–43
Swindale NV, Shoham D, Grinvald A, Bonhoeffer T, Hubener M (2000) Visual cortex maps are optimized for uniform coverage. Nat Neurosci 3:822–826
Tao H, Zhang LI, Bi G, Poo M-M (2000) Selective presynaptic propagation of long-term potentiation in defined neural networks. J Neurosci 20:3233–3243
Ts'o DY, Frostig RD, Lieke EE, Grinvald A (1990) Functional organization of primate visual cortex revealed by high resolution optical imaging. Science 249:417–420
von Melchner L, Pallas SL, Sur M (2000) Visual behaviour mediated by retinal projections directed to the auditory pathway. Nature 404:871–876
White LE, Bosking WH, Fitzpatrick D (2001) Consistent mapping of orientation preference across irregular functional domains in ferret visual cortex. Vis Neurosci 18:65–76
Willott JF (ed) (1983) The auditory psychobiology of the mouse. Thomas, Springfield, IL

Regional Forebrain Patterning and Neural Subtype Specification: Implications for Cerebral Cortical Functional Connectivity and the Pathogenesis of Neurodegenerative Diseases

MARK F. MEHLER[1]

1 Introduction

During the early embryonic period, dorsal telencephalic progenitor cells within the cortical ventricular zone (VZ) undergo radial migration and give rise to excitatory (glutamatergic) projection neurons that comprise the majority of cerebral cortical neurons and contribute to regional morphogenesis. By contrast, during perinatal life, current developmental models suggest that cortical oligodendrocytes (OLs) and astrocytes are generated from multipotent progenitors present within the cortical subventricular zone (SVZ). However, recent studies have shown that progenitor cells within the medial and lateral ganglionic eminences (MGE, LGE) of the ventral telencephalon can give rise to OLs and are also known to generate GABAergic neurons following tangential cortical migration. Although the majority of these ventral telencephalic progenitors disperse directly into the neocortex after migration, a subset of these LGE progenitors are destined for the cortical SVZ, where they give rise to both neuronal progenitors and to additional cells that become mitotically active within the cortical SVZ during the period of perinatal gliogenesis. Preliminary observations suggest that these cortical SVZ cells are derived from progenitors of the ganglionic eminences that are capable of generating both OLs and GABAergic neurons. These experimental findings are consistent with prior observations that OLs are generated from restricted ventral progenitor domains that also give rise to specific neuronal subtypes under the influence of graded levels of sonic hedgehog (Shh) signaling. In addition, there is experimental evidence that the sequential elaboration of neurons and OLs from these ventral telencephalic progenitors is potentiated by bone morphogenetic proteins (BMPs) and the BMP antagonist, noggin, respectively. BMPs and noggin are sequentially expressed within the cerebral cortex during the

[1] F.M. Kirby Program in Neural Protection and Repair, Departments of Neurology, Neuroscience and Psychiatry, Rose F. Kennedy Center for Research in Mental Retardation and Developmental Disabilities, Einstein Comprehensive Cancer Center, Albert Einstein College of Medicine, 1410 Pelham Parkway South, Bronx, New York 10461, USA

Results and Problems in Cell Differentiation, Vol. 39
C.F. Hohmann (Ed.): Cortical Development

periods of neurogenesis and gliogenesis, respectively. Shh has been shown to positively modulate the expression of two members (Olig1, 2) of a newly identified bHLH transcription factor subfamily that have been implicated in the earliest stages of OL lineage development. Additional studies reveal that Olig proteins exhibit dynamic profiles of regional and cellular expression within the embryonic telencephalon, and selective modulation by BMP signals within the dorsal forebrain. These overall observations suggest that specific ventral telencephalic progenitor cells of the ganglionic eminences undergo tangential migration to the cortical SVZ and sequentially give rise to distinct cortical GABAergic interneuron subtypes and OLs by local modulation of BMP signals. Further, Olig1 and 2 display distinct complementary and cooperative roles during cerebral cortical neurogenesis and gliogenesis.

These cumulative experimental findings suggest that Olig1 is involved in promoting OL lineage commitment, whereas Olig2 regulates a broader range of integrative developmental functions, including promoting neuronal-OL lineage restriction of self-renewing multipotent progenitors, actively inhibiting astrocyte lineage elaboration, modulating progenitor cell proliferation, enhancing and coordinately regulating neuronal and OL lineage commitment and preventing premature neuronal and glial terminal differentiation. Thus, specific regional environmental cues can orchestrate cortical fate decisions by modulating the expression of Olig1 and 2 within developmental stage-specific progenitor species. These experimental findings suggest an alternate model of cortical GABAergic neuronal and OL development that would incorporate the sequential contributions of the ventral and dorsal telencephalon to lineage specification and the known interdependence between evolving neuronal and OL species.

Tangential cortical migration furnishes specific neuronal and glial species that contribute to terminal cerebral cortical sculpting by modulating the activities of intrinsic cortical elements (glutamatergic neurons) initially derived by radial migration from the cortical VZ. These discrete developmental events occur within the broader context of regional cerebral cortical organizing centers that utilize released soluble signaling molecules to activate cascades of transcriptional regulators that initially establish unique progenitor cell domains and contribute to neural subtype specification and the integrity of evolving neural network connections. Neural subtype specification is orchestrated, in part, by combinatorial profiles of transcriptional regulators that endow individual regional neuronal subtypes with unique cellular signatures. An additional level of developmental regulation is provided by the changing and cell type-specific profiles of intercellular communications mediated by individual gap junction (connexin) proteins. Connexins are involved in regulating multipotent progenitor cell lineage restriction, neuronal and glial lineage commitment and progressive cellular differentiation. An additional level of regulatory control is provided by specific "neuronal" connexins that are preferentially expressed on specific regional cortical neuronal subtypes, including discrete GABAergic interneuron species. These inhibitory neuronal species contain electrical synapses that facilitate precisely synchronous spiking, the control of cortical activation and the emergent elaboration of complex cognitive and behavioral functions by functionally linking widely dispersed regional neuronal ensembles.

The molecular and physiological mechanisms underlying forebrain morphogenesis and neural subtype specification also have important implications for our understanding of the cause and ultimate prevention of specific classes of neuropsychiatric disorders. Neurodegenerative diseases are characterized clinically by the presence of severe and progressive cognitive, behavioral and associated neurological deficits, and pathologically by late-onset cell death of specific regional neuronal subtypes. Recent evidence suggests that these diseases of the aging brain may be caused by subtle and progressive developmental abnormalities in neural patterning and neuronal subtype specification that enhance the vulnerability of targeted regional neuronal subpopulations to normally non-lethal environmental stressors. These observations suggest that individual neurodegenerative disease pathogenic mutations may alter the cellular homeostasis of evolving neural stem and progenitor species during the stages of neural induction, neural tube patterning and regional neurogenesis. These pathological processes may contribute to the evolution of intermediate "mutant" cellular phenotypes that may be identifiable using sophisticated neuroimaging techniques and amenable to therapeutic interventions using innovative pharmacological, gene and stem cell approaches. The regional neural stem and progenitor cells that give rise to the specific primary neuronal subtypes that undergo irrevocable cellular injury and death in Alzheimer's and Huntington's disease are located in the MGE and the LGE, respectively. These regional uncommitted progenitor cells also give rise to additional neuronal subtypes that undergo radial and tangential migration and subsequent cellular maturation within areas of the basal forebrain and the cerebral cortex. In these and other neurodegenerative diseases, complex, progressive anatomically dispersed foci of cell death occur within discrete regional neuronal subpopulations particularly within the cerebral cortex. Therefore, in this chapter, we will elaborate on the theme that further examination of the normal molecular mechanisms regulating neural cell diversification within the developing forebrain may furnish fundamental insights into the pathogenesis of neurodegenerative diseases, the antemortem identification of the mutant cellular phenotype within vulnerable regional neural subpopulations and the development of novel therapeutic approaches to combat these pervasive, debilitating and notoriously difficult to treat chronic neurological disorders.

2
Role of the Ventral Telencephalon in Cerebral Cortical Development

Within the developing cerebral cortex of the dorsal telencephalon, neurons and radial glia are generated predominantly from early embryonic VZ self-renewing multipotent progenitors, whereas OLs and astrocytes are generated from similar late embryonic and perinatal SVZ progenitors (Cameron and Rakic 1991; Morshead et al. 1994; Gage et al. 1995; Weiss et al. 1996a,b; McKay 1997; Goldman et al. 1997). The majority of cortical neurons, the excitatory glutamatergic projection neurons, are generated in concert with radial glia during radial migration from the VZ (Parnavelas 2000; Noctor et al. 2001). In contrast, recent studies demonstrate that most if not all of the cortical GABAergic inhibitory interneurons are generated from progenitor cells that arise from the MGE and the LGE of the ventral telencephalon and undergo tangential cortical migration (de Carlos et al. 1996; Anderson et al. 1997, 2001; Tamamaki et al. 1997, 1999; Chapouton et al. 1999; Lavdas et al. 1999; Sussel et al. 1999; Wichterle et al. 1999). Progenitors from the MGE undergo discrete early and late waves of embryonic tangential migration and rapid cortical dispersion and elaboration of regional GABAergic interneurons (Anderson et al. 2001). By contrast, LGE progenitors only engage in late wave tangential cortical migration (Anderson et al. 2001). In addition, many of these LGE progenitors are retained within the cortical SVZ, where they give rise to GABAergic neurons and additional cells that become mitotically active at the time of cortical OL lineage elaboration (Anderson et al. 2001). Current developmental models suggest that cortical OLs are generated from multipotent progenitors present within the perinatal cortical SVZ (Levison and Goldman 1997). However, genetic and developmental studies suggest that cortical OLs can also arise from early embryonic multipotent progenitors of the ganglionic eminences of the ventral telencephalon (Martinez and Cobos 1998). This ventral origin of cortical OLs would be most compatible with observations from more caudal CNS regions where OLs are generated from restricted progenitor domains that also give rise to specific neuronal subtypes under the influence of Shh (Hardy and Friedrich 1996; Miller 1996; Richardson et al. 1997; Briscoe and Ericson 1999; Orentas et al. 1999; Briscoe et al. 2000).

Cortical GABAergic neurons that originate from the ganglionic eminences can be identified by the expression of Dlx1 and 2, and cells that originate from the MGE can be further defined by the expression of Lhx6 (Parnavelas 2000). These transcription factors are components of a combinatorial code for the elaboration of cortical GABAergic neurons that includes the ventral telencephalic basic helix-loop-helix (bHLH) factor, Mash1 (Lo et al. 1991; Casarosa et al. 1999; Parnavelas 2000). MGE and LGE progenitors give rise to different subtypes of cerebral cortical GABAergic neurons, defined by the expression of parvalbumin (PV), somatostatin (SS) or calretinin (CR), and additional neuronal subtypes that populate distinct CNS regions such as the striatum and the basal magnocellularis complex (Gonchar and Burkhalter 1997; Lavdas et al. 1999; Sussel et al. 1999; Marin et al. 2000; Parnavelas 2000; Noctor et al. 2001). These diverse neuronal subtypes have distinct trophic factor requirements (Pappas and Parnavelas 1998; Perez-Navarro et al. 2000). Within the evolving basal ganglia, progenitor cells that undergo tangential migration from MGE and LGE generative zones give rise to local interneurons, whereas progenitors that undergo radial migration give rise to projection neurons (Tamamaki et al. 1997; Parnavelas 2000; Wilson and Rubenstein 2000; Anderson et al. 2001). Projection neurons within each region maintain positional information necessary for the formation of topographic connectivity maps. Cholinergic neurons are generated exclusively from MGE progenitors that furnish interneurons to the developing striatum through tangential migration and projection neurons to the basal magnocellularis complex through radial migration. GABAergic projection neurons of the striatum and cholinergic projection neurons of the basal forebrain preferentially undergo neurodegeneration in Huntington's and Alzheimer's disease, respectively (Vonsattel et al. 1985; Engelborghs and DeDeyn 1997; Hof 1997; Braak and Braak 1998; Mitchell et al. 1999; Reddy et al. 1999; Glass et al. 2000; Perl 2000; Sieradzan and Mann 2001).

3 Developmental Actions of Neurogenic bHLH Genes

Neurogenesis is promoted by bHLH neural determination (proneural) and differentiation effector (neurogenic) genes (Anderson 1999). There are two major families of mammalian neural determination factors that are homologues of the Drosophila achaete-scute (i.e., Mash1) and atonal (i.e., Neurogenin [Ngn]1,2, Math3) gene families, respectively. Within the mammalian CNS, individual "proneural" genes are expressed in distinct neuronal subtypes (Anderson 1999). However, proneural genes from different bHLH families can also be expressed within the same neuronal cell type (i.e., Mash1, Math3; Tomita et al. 2000). Within the developing telencephalon, Ngn1 and Ngn2 are expressed in dorsal cell types, whereas Mash1 is expressed in ventral cell types (Anderson 1999). Mash1 can induce GABAergic neuronal traits when ectopically expressed in Ngn2-containing cells (Fode et al. 2000). There is increasing evidence that proneural genes mediate an ever-expanding range of developmental functions. Proneural genes regulate the transition from a self-renewing stem cell state to a more lineage-restricted neuronal-glial progenitor, progenitor cell proliferation, the acquisition of the neuronal fate, the coupling of neuronal lineage commitment to subtype specification and the active inhibition of astroglial lineage elaboration (Anderson 1999; Tomita et al. 2000; Nieto et al. 2001). The inhibition of the astroglial lineage by proneural genes is mediated, in part, by the active sequestration of complexes of transcription factors (i.e., Smad1) and transcriptional co-regulators (i.e., CBP/p300) from the promoters of astroglial differentiation genes and by inhibition of signaling molecules (i.e., STAT) required by cytokines (i.e., LIF) that promote astroglial lineage commitment (Sun et al. 2001). Downregulation of proneural gene expression at the time of cessation of neurogenesis reverses these molecular events, and allows cytokines that utilize the sequestered transcriptional complexes (i.e., BMPs/Smad1) to shift their developmental roles from neurogenic effectors to co-regulators of gliogenesis (Nakashima et al. 1999).

4
Mechanisms Regulating the Transition from Neurogenesis to Gliogenesis

The molecular mechanisms that regulate the transition from neurogenesis to gliogenesis are still poorly understood. Elements of the lateral inhibition pathway (Notch, Hes1 and 5, hesr2) promote astroglial lineage elaboration at the expense of neurogenesis (Hojo et al. 2000; Satow et al. 2001; Tanigaki et al. 2001). Activated Notch inhibits the generation of neurons from embryonic telencephalic progenitors and promotes the delayed emergence of glial cells from cortical SVZ progenitors (Chambers et al. 2001). Fibroblast growth factors (FGF) 1 and 2 are involved in the Notch-mediated inhibition of the neuronal lineage through selective upregulation of Notch expression and downregulation of Delta1 expression (Faux et al. 2001). Presenilin1, PS1, a gene that is mutated in certain familial, early-onset forms of Alzheimer's disease, is involved in mediating this developmental inhibition of neuronal lineage elaboration (Faux et al. 2001). In this context, PS1 can also regulate neurogenesis by modulating the balance of pre-patterning genes and the regional expression of Shh (Paganelli et al. 2001). These developmental actions are associated with the initial inhibition and the delayed promotion of progenitor cell proliferation. Transient Notch activation also promotes astroglial lineage commitment from adult NSCs by an instructive mechanism that is independent of the actions of gliogenic cytokines (i.e., BMPs, LIF; Morrison et al. 2000). Double but not single knockouts of members of the two families of proneural genes result in severe reductions in neurogenesis and precocious astroglial lineage elaboration (Tomita et al. 2000; Nieto et al. 2001).

Interestingly, the Mash1/Math3 double knockout is associated with the induction of hesr2 (Tomita et al. 2000). Math3 has been implicated in the regulation of developmental events downstream of Mash1, including progenitor cell cycle exit and early cellular migration from generative zones (Roztocil et al. 1997; Anderson 1999). However, the molecular mechanisms regulating the elaboration of the OL lineage, and the role of the lateral inhibition pathway in mediating this process, are presently unknown. A significant advance in our understanding of the developmental regulation of OL lineage elaboration has been the identification of the first genes encoding putative OL lineage-associated bHLH factors (Olig1 and 2; Lu et al. 2000; Zhou et al. 2000; Qi et al. 2001; Tekki-Kessaris et al. 2001; Woodruff et al. 2001). Olig1 and Olig2 have been implicated in the earliest stages of OL development. Interestingly, these factors comprise a distinct bHLH subfamily that displays greater homology to the Drosophila atonal than to the achaete-scute gene family (Lu et al. 2000; Zhou et al. 2000). Additional intrinsic developmental signals, including Mash1, can also regulate the timing of OL progenitor cell differentiation (Kondo and Raff 2000). Activation of Notch signaling by distinct Notch ligands expressed on the axons of developing neuroblasts and also on mature post-mitotic OLs can inhibit the differentiation of OL progenitors (Mitchell et al. 1999; Bongarzone et al. 2000; Wang and Barres 2000). These observations suggest that the two major families of neural bHLH genes have diverse cellular roles within individual progenitor species, and that the transition from neurogenesis to gliogenesis involves the reactivation of elements of the lateral inhibition pathway coincident with the downregulation of proneural bHLH genes.

5 Olig Genes and Regional Shh Signaling

The Olig genes are expressed in specific regional progenitor cell domains, including the ventral spinal cord, that give rise to OL precursors and specific ventral neuronal subtypes (Lu et al. 2000; Zhou et al. 2000). In addition, the sequential elaboration of Olig2 and Olig1 within the spinal cord precedes the expression of the earliest known OL precursor marker, platelet-derived growth factor α-receptor (PDGFαR) by 24–36 h (Zhou et al. 2000). Olig genes are expressed during the ventral to dorsal migration of spinal cord OL precursor cells from the VZ (Zhou et al. 2000). At the cellular level, the Olig genes are expressed in early OL lineage species but not in astrocytes (Lu et al. 2000). Unlike classical neural bHLH genes, the Olig genes continue to be expressed in the adult CNS (Lu et al. 2000). The potential complexity of Olig gene function has become apparent from several experimental observations. Ectopic expression of Olig1 and/or Olig2 in specific developmental contexts is sometimes insufficient to induce robust expression of OL lineage species (Lu et al. 2000; Zhou et al. 2000). These observations suggest that the Olig genes may require additional developmental stage-specific environmental signals or cellular co-factors for promotion of OL lineage elaboration (Zhou et al. 2000). In addition, the regional progenitor domains of Olig2 are considerable larger than those of Olig1, overlap with those of Mash1, and expression of Olig2 occurs at an earlier developmental stage than expression of Olig1 (Lu et al. 2000; Takebayashi et al. 2000; Zhou et al. 2000). Further, Olig2 is expressed in β-tubulin-immunoreactive nascent neuroblasts (Takebayashi et al. 2000). This is somewhat analogous to the expression of Mash1 in OL progenitors (Kondo and Raff 2000). Finally, in the early embryonic telencephalon, Olig2 but not Olig1 is expressed within ventral progenitor domains of the MGE and the LGE but not in the corresponding cerebral cortex (Takebayashi et al. 2000). These experimental findings suggest that Olig2 is involved in mediating an earlier and a broader set of developmental functions than Olig1, and may be involved in regulating the elaboration of both the neuronal and the OL lineages.

Throughout the neuraxis, Olig genes are initially expressed within regional ventral progenitor cell domains specified by concentrations of Shh required for the elaboration of specific neuronal subtypes and OLs (Lu et al. 2000; Takebayashi et al. 2000; Zhou et al. 2000). Shh is initially expressed in the anterior primitive streak and organizer at the gastrula stage and later within individual ventral telencephalic cells after specification of the MGE and LGE

(Gunhaga et al. 2000). This dual wave of Shh signaling may partially account for the analogous early and later profiles of expression of the Olig genes (Lu et al. 2000; Takebayashi et al. 2000; Zhou et al. 2000). The sequential modulation of Olig expression by cytokines may be required for mediating distinct sets of developmental functions. Both in vivo and in vitro, Shh positively regulates the expression of Olig genes and the sequential elaboration of progressive OL lineage species (Kohtz et al. 1998; Rowitch et al. 1999; Nery et al. 2001). Mash1 expression can also be induced by Shh signaling (Nery et al. 2001). However, OL lineage species can still be elaborated in the absence of endogenous Shh signaling (Nery et al. 2001). These observations suggest the possibility of alternate OL developmental lineages and regional signaling pathways. The existence of OL sublineages has been postulated on the basis of distinct profiles of platelet-derived growth factor (PDGF) α-receptor and proteolipid protein (PLP)/DM20 expression in regional progenitor species but has not been definitively proven (Thomas et al. 2000). Cytokines that regulate the proliferation, survival and differentiation of self-renewing multipotent progenitors, such as bFGF, can induce the expression of Gli, the Shh-signaling molecule (Brewster et al. 2000). Shh, in turn, may promote NSC lineage restriction and the transition to a program of progressive neuronal and OL lineage elaboration by positive modulation of bHLH factors such as Mash1 and Olig2 (Rowitch et al. 1999; Nery et al. 2001). We have shown that multipotent progenitors of the ventral telencephalon are selectively responsive to bFGF. Under the influence of Shh, these progenitors become more lineage-restricted neuronal/OL precursor species associated with the downregulation of nestin, the continued expression of the RNA-binding protein, Musashi (Kaneko et al. 2000), and the selective upregulation of Olig2 and Mash1.

Shh can also promote cellular proliferation through activation of G1 cyclins and engagement of the G1 cyclin-retinoblastoma susceptibility gene axis (Kenney and Rowitch 2000). Interestingly, N-terminal forms of Gli3 that are transcriptional repressors are preferentially bound to Smads, the BMP intracellular signaling molecules (Liu et al. 1998). The presence of BMPs induces a dissociation of the Gli3/Smad complex, subsequent Smad-mediated activation of BMP-responsive genes and Gli3-mediated antagonism of Shh activator functions (Liu et al. 1998). Modulation of BMP signals (see below) is responsible for the sequential elaboration of specific neuronal subtypes and OLs from these lineage-restricted ventral telencephalic progenitor species. Early proliferative neuroblasts (β-tubulin + /BrdU+) also continue to express Olig2 and Mash1 but not Olig1. By contrast, OL precursors and OL progenitor cells express Olig1, 2 and Mash1. Conversely, astroglia and differentiated, post-mitotic neurons and OLs do not express Olig1, 2 or Mash1. These cellular expression studies and additional Olig gene-manipulation paradigms suggest that Olig1 and Olig2 mediate a distinct spectrum of complementary and cooperative functions. Olig1 is predominantly involved in promoting OL lineage commitment and in preserving the integrity of OL progenitor species. By contrast, Olig2 potentiates neuronal-OL lineage restriction of self-renewing multipotent progenitors with concurrent active suppression of the astroglial lineage, enhances neuronal lineage commitment and prevents premature neuronal terminal differentiation. Olig1 and Olig2 also exhibit cooperative interactions in regulating the timing of OL lineage elaboration, and in preventing premature OL terminal differentiation. Mash1 has the potential to exhibit cooperative interactions with Olig2 (and to a lesser extent with Olig1) to help mediate these seminal developmental processes. In this regard, modulation of BMP signaling can dynamically regulate the regional cellular expression of Olig1 and Olig2.

6 Importance of Regional Forebrain Patterning for Neural Subtype Specification

There is increasing evidence that patterning of the embryonic cerebral cortex is a progressive developmental process that occurs over successive generative iterations (Wilson and Rubenstein 2000; Ragsdale and Grove 2001). Positional information is initially provided by soluble signaling molecules released from discrete organizing centers that establish regional structural and functional domains defined by the expression of specific profiles of transcription factors (Wilson and Rubenstein 2000; Ragsdale and Grove 2001). Transcriptional control of additional groups of downstream genes promotes later growth and more fine-grained regional specification. Thereafter, signals emanating from developing neural network connections help to orchestrate final regional structural and functional sculpting and the establishment of mature neural networks (Ragsdale and Grove 2001). Certain genes are expressed in invariable patterns during cortical development and in adult life (e.g., limbic system-associated membrane protein, LAMP), and serve as markers to define regional boundaries or specialized regional populations (Ragsdale and Grove 2001). By contrast, other profiles of developmental gene expression are transient (e.g., cadherins 6,8,11; inhibitor of differentiation 2) and define evolving regional boundaries, including microdomains within cortical generative zones. Some patterns of cortical gene expression are established prior to thalamocortical afferent innervation (e.g., LAMP, latexin), and appear to place constraints on when intrinsic regional patterns will be established (Ragsdale and Grove 2001). Further, early patterned cortical regions elaborate specific developmental trophic cues that determine the topography of afferent ingrowth. Distinct regional signaling centers (expressing BMPs, Wnts, Shh, FGFs) exist that contribute to progressive patterning along the anteroposterior, dorsoventral and mediolateral axes of the forebrain (Wilson and Rubenstein 2000; Ragsdale and Grove 2001). Regional organization involves both fate specification and regulated expansion of diverse regional progenitor species. There is extensive evidence that domains of the early evolving cerebral cortex compete for regional territories, and the diversity and specificity of cellular processing events unfolding within individual areas may be the primary impetus for final cortical pattern generation (Ragsdale and Grove 2001). Thus, early cortical patterning defects frequently result in later alterations in cellular expansion, migration and terminal differentiation (Ragsdale and Grove 2001).

Early cerebral cortical patterning can also be extensively modified by thalamocortical afferent innervation; these developmental interrelationships may be complex and multifactorial (Wilson and Rubenstein 2000; Ragsdale and Grove 2001). The interpretation of Shh signaling gradients is orchestrated by homeodomain transcriptional codes (including contributions from Nkx2.1, Nkx2.2, Gsh2, Pax6) that help to initially define progenitor cell domains. These regional domains are subsequently refined and maintained by cross-repressive interactions between complementary groups of homeodomain proteins that are expressed within contiguous regions (Briscoe and Ericson 1999; Briscoe et al. 2000). These molecular interactions help to orchestrate evolving neuronal subtype identity during cell cycle exit and terminal differentiation. Combinatorial profiles of homeodomain proteins determine the relative position of regional neuronal subtypes and regulate the activities of neural determination genes (Briscoe and Ericson 1999; Briscoe et al. 2000). For example, the homeodomain transcription factor Nkx 2.2 is involved in both neuronal and OL lineage specification and is expressed throughout the neuraxis (Briscoe et al. 1999; Xu et al. 2000; Qi et al. 2001). Within a specific progenitor cell domain, Nkx 2.2 actively promotes the induction of appropriate OL and associated neuronal subtype genes while simultaneously repressing the induction of non-sanctioned neuronal subtype genes that are normally expressed in adjacent progenitor domains (Briscoe and Ericson 1999; Briscoe et al. 2000). In addition, there are parallel sets of developmental cues that promote the expression of specific neurotransmitter or neuropeptide phenotypes, and other classes of maturational signals that link these general neuronal subclass traits to specific cellular processes and to individual domains within the neuraxis (Sakurada et al. 1999). Further diversification of neuronal subtypes required for the establishment of functional neural circuits is orchestrated by the expression of diverse classes of transcription factors and by additional guidance cues that promote discrete forms of axonal pathfinding and fasciculation (Ragsdale and Grove 2001).

Reciprocal interactions between sets of transcription factors (Pax6, Gsh2) mediate dorsoventral patterning and contribute to the elaboration of discrete progenitor cell domains on either side of the pallidal/subpallidal boundary (Wilson and Rubenstein 2000). FGF pathways regulate the earliest patterning of the ventral telencephalon, activate downstream gene products, such as Bf1, that control progenitor cell proliferation and initial regional specification, and establish a developmental framework of progenitor cell responsiveness for refinement of telencephalic boundaries by regional gradient morphogens. Shh is expressed at maximal levels at the midline of the ventral forebrain, and a mediolateral concentration gradient is thereafter established that contributes to the specification of the MGE and the LGE, and the induction of region-specific transcription factors that help to sculpt the identities of domain-specific progenitor species (Wilson and Rubenstein 2000; Anderson et al. 2001).

Nkx2.1 is expressed exclusively in the MGE, whereas Gsh2 is present in both the MGE and the LGE (Wilson and Rubenstein 2000; Anderson et al. 2001). The opposing actions of Pax6 and Emx2 establish positional identities within the dorsal telencephalon and contribute to the elaboration of a homeodomain code to regulate the establishment of telencephalic subdivisions (Ragdale and Grove 2001). Within these evolving dorsal forebrain regions, BMP signaling represses the expression of ventral forebrain markers of progenitor cell domains (e.g., Nkx2.1) and mediators of tangential cortical migration (e.g., Dlx2). Another important marker of Shh-mediated ventral progenitor cell specification is the Nkx2.2 transcription factor (Briscoe et al. 1999; Xu et al. 2000; Qi et al. 2001). OLs within the forebrain are generated from Nkx2.2-expressing neuroepithelial cells that undergo subsequent migration and sequential expression of increasingly mature OL lineage markers identified using PDGFαR, O4 and Ranscht monoclonal antibodies (Qi et al. 2001). Nkx2.2 is downregulated coincident with OL progenitor exit from cell cycle and terminal differentiation, heralded by the expression of an epitope of galactocerebroside (GC) identified by use of the GC/O1 monoclonal antibody (Qi et al. 2001). Studies employing Nkx2.1 and Nkx2.2 knockout mice suggest that both of these transcriptional regulators contribute to the elaboration of forebrain OLs (Qi et al. 2001). However, there is preliminary evidence that these distinct Shh-regulated transcription factors may be involved in the generation of separate regional telencephalic OL populations that may be associated with distinct pallidal and subpallidal neuronal subtypes. Nkx2.2 regulates OL differentiation but not initial OL progenitor specification (Qi et al. 2001). This experimental observation is consistent with the bimodal pattern of expression of Nkx2.2 in uncommitted progenitors and in later OL precursor species. These findings also support accumulating evidence that elements of regional and cell-type restriction are already present within very early forebrain progenitor populations. Further, they suggest that extrinsic regulatory cues may combine with or initiate the expression and activation of intrinsic determinants of lineal restriction within the developing forebrain.

7
Role of Local BMP Signaling in Cerebral Cortical Neuronal and OL Lineage Elaboration

We have previously shown that BMPs inhibit lineage commitment of early-born cerebral cortical neurons and negatively regulate VZ cell number by promoting apoptosis (Mabie et al. 1999). By contrast, the migration and differentiation of these early-born neurons from the cortical VZ is promoted by BMP ligands (Li et al. 1998). Coincident with the establishment of the cortical SVZ, BMPs then promote the lineage commitment of late-born cortical neurons and inhibit OL lineage elaboration from cortical SVZ progenitors (Mabie et al. 1999). Exposure of SVZ progenitor cells to the BMP antagonist noggin promotes the generation of OLs and inhibits further neuronal lineage elaboration (Mabie et al. 1999). In preliminary studies, we have also demonstrated that application of BMPs to embryonic MGE and LGE progenitor cells selectively enhances the generation of GABAergic neurons, whereas exposure to noggin enhances the elaboration of OLs at the expense of these neurons. BMPs and noggin are sequentially expressed within the cerebral cortex during the periods of neurogenesis and gliogenesis, respectively (Furuta et al. 1997; Ebendal et al. 1998). Recent studies suggest that differences in the cellular effects of BMP signaling are mediated by changing profiles of BMP receptor (BMPR) expression on target progenitor species (Panchision et al. 2001). For example, there is experimental evidence that BMPR1A modulates early progenitor cell expansion and imparts instructive neural lineage cues. By contrast, BMPR1A induces the expression of BMPR1B, an additional signaling subunit that promotes mitotic arrest and context-dependent early developmental apoptosis and later enhancement of terminal differentiation. These BMPR signaling subunits exist as part of a more extensive network of additional modulatory cues that regulate the timing and duration of BMP transduction events and associated stage-specific developmental processes (Panchision et al. 2001). By contrast, our preliminary observations reveal that LIF promotes the lineage commitment of both early-born neurons and radial glia and concurrently inhibits the generation of OL precursors from embryonic cerebral cortical progenitor cells at a developmental stage prior to the time of tangential cortical migration of those ventral telencephalic precursor cells that give rise to cortical GABAergic interneurons (Mehler et al. 2000). These overall observations suggest that, following tangential migration to the cortical SVZ, ventral telencephalic progenitor cells of the ganglionic eminences sequentially give rise to specific GABAergic interneuron subtypes and OLs by sequential local environmental modulation of BMP signals and differential regulation of Olig 1 and Olig2 with developmental stage-specific progenitor species. Therefore, regional dorsoventral patterning of the embryonic forebrain is mediated by the dynamic interplay between specific classes of soluble signals and gradient morphogens, and by alterations in the stoichiometry within target progenitor species of their downstream transcriptional effectors (Lumsden and Krumlauf 1996; Liu et al. 1998).

8
Generation of OL Lineage Species in the Adult Brain: Therapeutic Implications

During progressive stages of cerebral cortical OL development, bipolar PDGFα receptor and NG2 proteoglycan expressing OL precursors migrate from the cortical SVZ to the subcortical white matter and the neocortex (Pfeiffer et al. 1993; Miller et al. 1999). The majority of these OL precursor species rapidly evolve into committed O4-expressing OL progenitors and undergo cell cycle exit and progressive cellular differentiation into post-mitotic (GC/O1-

immunoreactive) OLs that express various myelin proteins (Pfeiffer et al. 1993; Miller et al. 1999). However, a small proportion of these OL precursors remain as reactive intermediate progenitor pools during later postnatal and adult life (Levine et al. 1993; Nishiyama et al. 1997, 1999; Reynolds and Hardy 1997; Redwine and Armstrong 1998). Within the adult CNS, these cells comprise 70% of proliferating cells and 5–8% of the total glial population (Levine et al. 2001). These OL precursor populations and additional self-renewing multipotent progenitors of the residual adult subependymal zone may be activated in response to specific forms of CNS injuries, including those that induce widespread inflammation and demyelination (Craig et al. 1996; Franklin et al. 1997; Carroll et al. 1998; Keirstead et al. 1998; Magavi et al. 2000). These two distinct progenitor populations may furnish myelinating OLs that can participate in effective forms of neural regenerative responses. The conversion of adult OL precursors into myelinating OLs in response to CNS injury may also forestall their production of chondroitin sulfate proteoglycans that can effectively block axonal regeneration (Fawcett and Asher 1999; Levine et al. 2001). These cumulative observations suggest that understanding the intrinsic and environmental regulation of early cerebral cortical OL lineage development will help to identify therapeutic targets for the application of innovative neural regenerative strategies.

9
Role of Gap Junction Channels and GABAergic Neuronal Subtypes in Cerebral Cortical Functional Connectivity

An additional level of cerebral cortical developmental regulation is provided by the changing profiles of intercellular communications mediated by different classes of gap junction channels (connexins) expressed between progenitor and also more differentiated neural lineage species (Rozental et al. 2000). Multipotent progenitors that express nestin are initially coupled by gap junction channels composed of connexin 43 (Nadarajah et al. 1998; Rozental et al. 2000). These early developmental species also express transcripts for connexin 45. Coincident with neuronal lineage commitment and the expression of early neurofilament proteins, progenitor cell coupling strength declines and there is downregulation of connexin 43 (Rozental et al. 2000). These changes in coupling efficiency are associated with the concurrent expression of connexins 26, 33, 36, 40 and 45. Terminal phases of neuronal differentiation are heralded by the expression of more mature neurofilament proteins, morphologic maturation, the elaboration of inward currents and action potentials, and further changes in the profiles of connexin expression, including an upregulation of connexin 36 (Rozental et al. 2000). Recent studies suggest that selected "neuronal" connexins such as connexin 36 and 47 are expressed on specific cerebral cortical neuronal subtypes and endow these cells with certain physiological properties that establish and subsequently define functional microdomains, and contribute to cortical connectivity and the emergent properties of cognitive and behavioral processing (Galarreta and Hestrin 2001). Cortical PV- and SS-containing GABAergic interneuron subtypes express connexin 36, whereas some cortical excitatory projection neurons express connexin 47. Connexin 36 knockout mice exhibit a near total absence of electrical synapses on cortical interneurons and a weaker than normal synchrony of cortical rhythms (Deans et al. 2001). These observations suggest that connexin 36 plays a significant role in the generation of widespread, synchronous inhibitory activities through the actions of electrical synapses.

Cortical inhibitory interneurons are extensively coupled by electrotonic synapses that define functionally diverse networks (Galarreta and Hestrin 2001). Fast-spiking (FS) cortical interneurons of layers II–VI represent a highly connected electrical network of PV-containing basket cells joined by the presence of gap junctions between dendrites. Another class of SS- but not PV-containing low-threshold spiking cortical (LTS) interneurons of layers IV–VI also exhibit extensive electrical synapse interconnections. In addition to the extensive coupling among FS or LTS interneurons, there are rare interconnections between these interneuron subtypes and also between LTS interneurons and excitatory projection neurons (Galarreta and Hestrin 2001). Further, the presence of both electrical and chemical synapses between cortical FS interneurons dramatically increases the integrative capacity of the inhibitory neuronal network (Galarreta and Hestrin 2001). Electrotonic synapses facilitate precisely synchronous spiking that contributes to the overall control of regional cortical activation. Additional signaling inputs arising from metabotropic glutamatergic receptor activation induce differential LTS and FS spike synchronization in a graded fashion (Deans et al. 2001; Galarreta and Hestrin 2001). Further, the temporal profiles of GABA-mediated inhibitory responses determine the frequency of neuronal oscillations. Finally, spike-to-spike transmission is very precisely regulated between excitatory and inhibitory neurons residing within the neocortex (Galarreta and Hestrin 2001). These overall physiological properties permit interneuronal networks connected by both electrotonic and chemical synapses to be particularly sensitive to synchronous relationships between excitatory cellular inputs. This facilitates the detection of coincidence and the establishment of other emergent properties that are required for facilitating complex cortical cognitive and behavioral processing (Galarreta and Hestrin 2001).

10
Regional Forebrain Patterning and Neurodegenerative Diseases

There is an increasing body of evidence that neurodegenerative diseases may represent fundamental disorders of neural patterning and neural subtype specification (Mehler and Gokhan 2000, 2001). Neurodegenerative disease pathogenic mutations have the potential to partially disrupt the interactions of the aberrant protein product with a series of protein-binding partners that are involved in neural induction, patterning of the neural tube and progressive stages of neurogenesis (Mehler and Gokhan 2001). The resulting progressive but subtle and selective developmental abnormalities can, in turn, alter the homeostasis of targeted regional neural stem and progenitor species and predispose their vulnerable neuronal and glial progeny to irrevocable cellular injury and death from normally non-lethal environmental cues (Mehler and Gokhan 2000, 2001). Although the molecular mechanisms governing the elaboration and maintenance of mature regional neural subtypes from neural stem and progenitor species is still largely undefined, the developmental interplay of gradient morphogens, cytokines and transcriptional regulators endows these evolving neural subpopulations with unique biochemical and physiological properties. The hallmark of neurodegenerative diseases is the occurrence of precise temporospatial gradients of adult-onset neural cell death that are disease-selective (Vonsattel et al. 1985; Engelborghs and DeDeyn 1997; Hof 1997; Braak and Braak 1998; Mitchell et al. 1999; Reddy et al. 1999; Glass et al. 2000; Perl 2000; Sieradzan and Mann 2001). The profiles of progressive cell loss cannot easily be explained by examining the intrinsic cellular properties of the often widely dispersed regional neural subpopulations involved or by defining common patterns of connectivity. However, various degrees of overlap in the profiles of affected developmental components comprising the combinatorial codes of transcription factors and other molecular signals that define diverse neural subtypes rendered differentially vulnerable by the pathogenic mutations represent an attractive mechanism to explain the unique patterns of late-onset regional cell loss seen in individual neurodegenerative diseases (Mehler and Gokhan 2000).

The regional neural stem and progenitor cells that give rise to the neuronal subtypes that undergo the earliest and most severe forms of neurodegeneration in Alzheimer's and Huntington's disease are located in the MGE and the LGE, respectively (Vonsattel et al. 1985; Engelborghs and DeDeyn 1997; Hof 1997; Braak and Braak 1998; Mitchell et al. 1999; Reddy et al. 1999; Glass et al. 2000; Parnavelas 2000; Perl 2000; Wilson and Rubenstein 2000; Anderson et al. 2001; Sieradzan and Mann 2001). In addition, areas of the cerebral cortex exhibit complex secondary gradients of neural cell death in both diseases (Vonsattel et al. 1985; Engelborghs and DeDeyn 1997; Hof 1997; Braak and Braak 1998; Mitchell et al. 1999; Reddy et al. 1999; Glass et al. 2000; Perl 2000; Sieradzan and Mann 2001). Recent evidence suggests that these anatomically dispersed pathological events occur in parallel (Reddy et al. 1999; Sieradzan and Mann 2001). Therefore, detailed comparative characterization of the physiological status of biosynthetic pathways of affected and unaffected regional ventral and dorsal forebrain neural subpopulations in wild-type and in mutant mice harboring the relevant pathogenic mutations will help to clarify and define the basic molecular pathology underlying these neurodegenerative diseases. These studies will also help to identify the precise cellular alterations present in progenitor and more differentiated neural species that exhibit an intermediate "mutant" cellular phenotype (Mehler and Gokhan 2000, 2001). The use of sophisticated functional neuroimaging techniques may permit the identification of this mutant intermediate state within developmental species present in the living brains of pre-symptomatic individuals at risk of acquiring these neurodegenerative diseases (Louie et al. 2000; Ross and Bluml 2001). An important outgrowth of these experimental observations is the potential for developing innovative therapeutic strategies to combat the pathological consequences of these neurodegenerative disease mutations using emerging pharmacological, gene and stem cell approaches prior to the occurrence of irrevocable cellular injury and death (Horner and Gage 2000). It may even be possible to identify the initial developmental populations at risk during early embryonic life prior to the establishment of an enduring intermediate mutant progenitor cell state. This would allow therapeutic interventions to commence in utero. At a fundamental biological level, these important investigations will determine why neural stem and progenitor cells residing within specific regions of the ventral forebrain give rise to regional neuronal subtypes that are so differentially affected by the disease process (Vonsattel et al. 1985; Engelborghs and DeDeyn 1997; Hof 1997; Braak and Braak 1998; Mitchell et al. 1999; Reddy et al. 1999; Glass et al. 2000; Parnavelas 2000; Perl 2000; Wilson and Rubenstein 2000; Anderson et al. 2001; Sieradzan and Mann 2001). These studies will help to clarify where within the normal progenitor cell biosynthetic cascade neural subtype specification occurs, and whether specific regional environmental cues endow these evolving progenitor species with significant degrees of developmental plasticity (McCarthy et al. 2001). Conversely, these explorations will also define whether significant heterogeneity exists in neural stem cell pools within regional generative zones of the ventral and the dorsal telencephalon. These cumulative investigations will establish a molecular blueprint for defining the profiles of mutant progenitor cell states that are responsible for generating the unique primary and secondary foci of regional late-onset cell death characteristic of individual neurodegenerative diseases.

11 Summary and Future Directions

Study of the role of forebrain patterning in neural subtype specification is still in its infancy. These investigations are part of an overall examination of the intricate interrelationships between neural induction, patterning of the neural tube and progressive stages of neurogenesis. Despite these caveats, considerable progress has already been made in defining the general organizing plan underlying cerebral cortical development, and the key sets of environmental cues, signaling pathways and transcriptional regulators that orchestrate the progressive elaboration of complex regional microdomains. Recent studies have suggested that neuronal and glial subtype specification, and the coordinate regulation and maintenance of the mature cellular phenotype are elaborate developmental events that progressively unfold between the time of neural induction and the eventual establishment of appropriate neural network connections. In addition, the establishment of specialized regional microdomains characteristic of the mature cerebral cortex appears to involve a dynamic interplay between gradient morphogens, their cascades of downstream transcriptional targets, local and long-range migratory waves of neural stem and progenitor species and additional local environmental signals that modulate and sculpt the molecular codes that are ultimately responsible for the fine-grained process of neural subtype specification. An additional level of developmental regulation is provided by targeted cell-cell communications mediated through the changing profiles of gap junction channels present on distinct progenitor cell and more mature lineage species.

Recent evidence suggests that specific gap junction (connexin) proteins are intimately involved in modulating cortical excitation and in establishing the biophysical cellular signatures required for coordinating the temporospatial-defined activation of neuronal ensembles that underlie integrative neural network and higher order cognitive and behavioral processing. Components of these elaborate forebrain networks are differentially affected by individual neurodegenerative diseases. These diverse neurological disorders are all characterized by a unique and currently undefined pathological mechanism that results in gradients of late-onset cell death of regional neuronal and glial subtypes. Numerous recent experimental observations suggest that the individual pathogenic mutations subtly compromise cumulative developmental processes involved in neural subtype specification, and establish an intermediate mutant cellular phenotype that predisposes these now vulnerable target populations to undergo irrevocable cellular injury and death in response to normally nonlethal environmental stressors.

These exciting findings suggest that it will be possible to identify these developmental cellular species at risk in presymptomatic individuals harboring the neurodegenerative gene mutations, and to prevent the initiation and progression of regional cell loss using a variety of innovative neuroprotective and neural regenerative strategies. Future studies will further refine our understanding of the normal molecular mechanisms underlying cerebral cortical development. These investigations will also clarify and define the extent of ongoing regional progenitor cell plasticity, the limits of local environmental influences and the regional timing of lineal development. These cumulative observations will be essential to allow us to define the pathogenetic basis of a broad range of diseases of the developing, mature and aging brain, and will pave the way for the application of emerging therapeutic strategies to reconstitute the individual cellular elements and the integrative neural networks adversely affected in specific classes of neurological disorders.

Acknowledgements. Experimental studies cited from the authors laboratory have been generously supported by the NINDS, NIH, the F.M. Kirby Foundation and the Alpern Family Foundation. I am grateful to Solen Gokhan and to Shau Yung for invaluable intellectual and scientific contributions.

References

Anderson DJ (1999) Lineages and transcription factors in the specification of vertebrate primary sensory neurons. Curr Opin Neurobiol 9:517–524

Anderson SA, Eisenstat DD, Shi L, Rubenstein JLR (1997) Interneuron migration from basal forebrain to neocortex: dependence on Dlx genes. Science 278:474–476

Anderson SA, Marin O, Horn C, Jennings K, Rubenstein JLR (2001) Distinct cortical migrations from the medial and lateral ganglionic eminences. Development 128:358–363

Birling MC, Price J (1998) A study of the potential of the embryonic rat telencephalon to generate oligodendrocytes. Dev Biol 193:110–113

Bongarzone ER, Byravan S, Givogri MI, Schonmann V, Campagnoni AT (2000) Platelet-derived growth factor and basic fibroblast growth factor regulate cell proliferation and the expression of notch-1 receptor in a new oligodendrocyte cell line. J Neurosci Res 62:319–328

Braak H, Braak E (1998) Evolution of neuronal changes in the course of Alzheimer's disease. J Neural Transm Suppl 53:127–140

Brewster R, Mullor JL, Ruiz I, Altaba A (2000) Gli2 functions in FGF signaling during antero-posterior patterning. Development 127:4395–4405

Briscoe J, Ericson J (1999) The specification of neuronal identity by graded sonic hedgehog signaling. Cell Dev Biol 10:353–362

Briscoe J, Sussel L, Serup P, Hartigan-O'Connor D, Jessel TM, Rubenstein JLR, Ericson J (1999) Homeobox gene NKx2.2 and specification of neuronal identity by graded sonic hedgehog signaling. Nature 398:622–627

Briscoe J, Pierani A, Jessell TM, Ericson J (2000) A homeodomain protein code specifies progenitor cell identity and neuronal fate in the ventral neural tube. Cell 101:435–445

Cameron RS, Rakic P (1991) Glial cell lineage in the cerebral cortex: a review and synthesis. GLIA 4:124–137

Carroll WM, Jennings AR, Ironside LJ (1998) Identification of the adult resting progenitor cell by autoradiographic tracking of oligodendrocyte precursors in experimental CNS demyelination. Brain 121:293–302

Casarosa S, Fode C, Guillemot F (1999) Mash1 regulates neurogenesis in the ventral telencephalon. Development 126 525–534

Chambers CB, Peng Y, Nguyen H, Gaiano N, Fishell G, Nye JS (2001) Spatiotemporal selectivity of response to Notch1 signals in mammalian forebrain precursors. Development 128:689–702

Chapouton P, Gartner A, Gotz M (1999) The role of Pax6 in restricting cell migration between developing cortex and basal ganglia. Development 126:5569–5579

Craig CG, Tropepe V, Morshead CM, Reynolds BA, Weiss S, van der Kooy D (1996) In vivo growth factor expansion of endogenous subependymal neural precursor cell populations in the adult mouse brain. J Neurosci 16:2649–2658

de Carlos JA, Lopez-Mascaraque L, Valverde F (1996) Dynamics of cell migration from the lateral ganglionic eminence in the rat. J Neurosci 16:6146–6156

Deans MR, Gibson JR, Sellitto C, Connors BW, Paul DL (2001) Synchronous activity of inhibitory networks in neocortex requires electrical synapses containing connexin 36. Neuron 31: 477–485

Ebendal T, Bengtsson H, Soderstrom S (1998) Bone morphogenetic proteins and their receptors: potential functions in the brain. J Neurosci Res 15:139–146

Engelborghs S, DeDeyn PP (1997) The neurochemistry of Alzheimer's disease. Acta Neurol Belg 97:67–84

Faux CH, Turnley AM, Epa R, Cappai R, Bartlett PF (2001) Interactions between fibroblast growth factors and notch regulate neuronal differentiation. J Neurosci 21:5587–5596

Fawcett JW, Asher RA (1999) The glial scar and CNS repair. Brain Res Bull 49:377–391

Fode C, Ma Q, Casarosa S, Ang SL, Anderson DJ, Guillemot F (2000) A role for neural determination genes in specifying the dorsoventral identity of telencephalic neurons. Genes Dev 14:67–80

Franklin RJ, Gilson JM, Blakemore WF (1997) Local recruitment of remyelinating cells in the repair of demyelination in the central nervous system. J Neurosci Res 50:337–344

Furuta Y, Piston DW, Hogan BLM (1997) Bone morphogenetic proteins (BMPs) as regulators of dorsal forebrain development. Development 124:2203–2212

Gage FH, Ray J, Fisher LJ (1995) Isolation, characterization and use of stem cells from the CNS. Annu Rev Neurosci 18:159–192

Galarreta M, Hestrin S (2001) Electrical synapses between GABA-releasing interneurons. Nat Rev 2:425–433

Glass M, Dragunow M, Faull RL (2000) The pattern of neurodegeneration in Huntington's disease: a comparative study of cannabinoid, dopamine, adenosine and GABA(A) receptor alterations in the human basal ganglia in Huntington's disease. Neuroscience 97:505–519

Goldman JE, Zerlin M, Newman S, Zhang L, Gensert J (1997) Fate determination and migration of progenitors in the postnatal mammalian CNS. Dev Neurosci 19:42–48

Gonchar Y, Burkhalter A (1997) Three distinct families of GABAergic neurons in rat visual cortex. Cerebral Cortex 7:347–358

Gunhaga L, Jessell TM, Edlund T (2000) Sonic hedgehog signaling at gastrula stages specifies ventral telencephalic cells in the chick embryo. Development 127:3283–3293

Hardy RJ, Friedrich VL (1996) Oligodendrocyte progenitors are generated throughout the embryonic mouse brain but differentiate in restricted foci. Development 122:2059–2069

Hof PR (1997) Morphology and neurochemical characteristics of the vulnerable neurons in brain aging and Alzheimer's disease. Eur Neurol 37:71–81

Hojo M, Ohtsuka T, Hashimoto N, Gradwohl G, Guillemot F, Kageyama R (2000) Glial cell fate specification modulated by the bHLH gene Hes5 in mouse retina. Development 127:2515–2522

Horner PJ, Gage FH (2000) Regenerating the damaged central nervous system. Nature 407: 963–970

Kaneko Y, Sakakibara S, Imai T, Suzuki A, Nakamura Y, Sawamoto K, Ogawa Y, Toyama Y, Miyata T, Okano H (2000) Musashi1: an evolutionally conserved marker for CNS progenitor cells including neural stem cells. Dev Neurosci 22:139–15

Keirstead HS, Levine JM, Blakemore WF (1998) Response of the oligodendrocyte progenitor cell population (defined by NG2 labeling) to demyelination of the adult spinal cord. GLIA 22: 161–170

Kenney AM, Rowitch DH (2000) Sonic hedgehog promotes G(1) cyclin expression and sustained cell cycle progression in mammalian neuronal precursors. Mol Cell Biol 20:9055–9067

Kohtz JD, Baker DP, Corte G, Fishell G (1998) Regionalization within the mammalian telencephalon is mediated by changes in responsiveness to sonic hedgehog. Development 125: 5079–5089

Kondo T, Raff M (2000) Basic helix-loop-helix proteins and the timing of oligodendrocyte differentiation. Development 127:2989–2998

Lavdas AA, Grigoriou M, Pachnis V, Parnavelas JG (1999) The medial ganglionic eminence gives rise to a population of early neurons in the developing cerebral cortex. J Neurosci 19:7881–7888

Levine JM, Stincone F, Lee YS (1993) Development and differentiation of glial precursor cells in the rat cerebellum. GLIA 7:307–321

Levine JM, Reynolds R, Fawcett JW (2001) The oligodendrocyte precursor cell in health and disease. Trends Neurosci 24:39–47

Levison SW, Goldman JE (1997) Multipotential and lineage restricted precursors coexist in the mammalian perinatal subventricular zone. J Neurosci Res 48:83–94

Li WW, Cogswell CA, LoTurco JJ (1998) Neuronal differentiation of precursors in the neocortical ventricular zone is triggered by BMP. J Neurosci 18:8853–8862

Liu F, Massague J, Ruiz i Altaba A (1998) Carboxy-terminally truncated Gli3 proteins associate with Smads. Nat Genet 20:325–326

Lo LC, Johnson JE, Wuenschell CW, Saito T, Anderson D J (1991) Mammalian *achaete-scute* homolog 1 is transiently expressed by spatially-restricted subsets of early neuroepithelial and neural crest cells. Genes Dev 5:1524–1537

Louie AY, HŸber MM, Ahrens ET, Rothbˇcher U, Moats R, Jacobs RE, Fraser SE, Meade TJ (2000) In vivo visualization of gene expression using magnetic resonance imaging. Nat Biotechnol 18:321–325
Lu QR, Yuk D, Alberta JA, Zhu Z, Pawlitzky I, Chan J, McMahon AP, Stiles CD, Rowitch DH (2000) Sonic hedgehog-regulated oligodendrocyte lineage genes encoding bHLH proteins in the mammalian central nervous system. Neuron 25:317–329
Lumsden A, Krumlauf R (1996) Patterning the vertebrate neuraxis. Science 274:1109–1115
Mabie PC, Mehler MF, Kessler JA (1999) Multiple roles of bone morphogenetic protein signaling in the regulation of cortical cell number and phenotype. J Neurosci 19:7077–7088
Magavi SS, Leavitt BR, Macklis JD (2000) Induction of neurogenesis in the neocortex of adult mice. Nature 405:951–955
Marin O, Anderson SA, Rubenstein JLR (2000) Origin and molecular specification of striatal interneurons. J Neurosci 20:6063–6076
Martinez S, Cobos J (1998) Regional and cellular specification in the prosencephalic anlage of the neural plate. Eur J Neurosci 10 [Suppl]:323
McCarthy M, Turnbull DH, Walsh CA, Fishell G (2001) Telencephalic neural progenitors appear to be restricted to regional and glial fates before the onset of neurogenesis. J Neurosci 21: 6772–6781
McKay R (1997) Stem cells in the central nervous system. Science 276:66–71
Mehler MF, Gokhan S (2000) Mechanisms underlying neural cell death in neurodegenerative diseases: alterations of a developmentally-mediated cellular rheostat. Trends Neurosci 23: 599–605
Mehler MF, Gokhan S (2001) Developmental mechanisms in the pathogenesis of neurodegenerative diseases. Prog Neurobiol 63:337–363
Mehler MF, Gokhan S, Mabie PC, Kessler JA (2000) Embryonic cerebral cortical astrocytes are generated from multipotent neural precursor pools under the influence of distinct environmental signals. Soc Neurosci Abstr 26:1069
Miller RH (1996) Oligodendrocyte origins. Trends Neurosci 19:92–96
Miller RH, Hayes JE, Dyer KL, Sussman CR (1999) Mechanisms of oligodendrocyte commitment in the vertebrate CNS. Int J Dev Neurosci 17:753–763
Mitchell IJ, Cooper AJ, Griffiths MR (1999) The selective vulnerability of striatopallidal neurons. Prog Neurobiol 59:691–719
Morrison SJ, Perez SE, Qiao Z, Verdi JM, Hicks C, Weinmaster G, Anderson DJ (2000) Transient Notch activation initiates an irreversible switch from neurogenesis to gliogenesis by neural crest stem cells. Cell 101:499–510
Morshead CM, Reynolds BA, Craig CG, McBurney MW, Staines WA, Morasutti D, Weiss S, van der Kooy D (1994) Neural stem cells in the adult mammalian forebrain: a relatively quiescent subpopulation of subependymal cells. Neuron 13:1071–1082
Nadarajah B, Makarenkova H, Becker DL, Evans WH, Parnavelas JG (1998) Basic FGF increases communication between cells of the developing neocortex. J Neurosci 18:7881–7890
Nakashima K, Yanagisawa M, Arakawa H, Kimura N, Hisatsune T, Kawabata M, Miyazono K, Taga T (1999) Synergistic signaling in fetal brain by STAT3-Smad1 complex bridged by p300. Science 284:479–482
Nery S, Wichterle H, Fishell G (2001) Sonic hedgehog contributes to oligodendrocyte specification in the mammalian forebrain. Development 128:527–540
Nieto M, Schurmans C, Britz O, Guillemot F (2001) Neural bHLH genes control the neuronal versus glial fate decision in cortical progenitors. Neuron 29:401–413
Nishiyama A, Yu M, Drazba JA, Tuohy VK (1997) Normal and reactive NG2+ glial cells are distinct from resting and activated microglia. J Neurosci Res 48:299–312
Nishiyama A, Chang A, Trapp BD (1999) NG2+ glial cells: a novel glial cell population in the adult brain. J Neuropathol Exp Neurol 58:1113–1124
Noctor SC, Flint AC, Weissman TA, Dammerman RS, Kriegstein AR (2001) Neurons derived from radial glial cells establish radial units in neocortex. Nature 409:714–720

Orentas DM, Hayes JE, Dyer KL, Miller RH (1999) Sonic hedgehog signaling is required during the appearance of spinal cord oligodendrocyte precursors. Development 126:2419–2429

Paganelli AR, Oscana OH, Prat MI, Franco PG, Lopez SL, Morelli L, Adamo AM, Riccomagno MM, Matsubara E, Shoji M, Affranchino JL, Castano EM, Carrasco AE (2001) The Alzheimer-related gene presenilin-1 facilitates sonic hedgehog expression in Xenopus primary neurogenesis. Mech Dev 107:119–131

Panchision DM, Pickel JM, Studer L, Lee S-H, Turner PA, Hazel TG, McKay RDG (2001) Sequential actions of BMP receptors control neural precursor cell production and fate. Genes Dev 15:2094–2110

Pappas IS, Parnavelas JG (1998) Basic fibroblast growth factor promotes the generation and differentiation of calretinin neurons in the rat cerebral cortex in vitro. Eur J Neurosci 10:1436–1445

Parnavelas JG (2000) The origin and migration of cortical neurones: new vistas. Trends Neurosci 23:126–131

Perez-Navarro E, Akerud P, Marco S, Canals JM, Tolosa E, Arenas E, Alberch J (2000) Neurturin protects striatal projection neurons but not interneurons in a rat model of Huntington's disease. Neuroscience 98:89–96

Perl DP (2000) Neuropathology of Alzheimer's disease and related disorders. Neurol Clin 18: 847–864

Pfeiffer SE, Warrington AE, Bansal R (1993) Oligodendrocyte and its many cellular processes. Trends Cell Biol 3:191–197

Qi K, Cal J, Wu Y, Rui W, Lee J, Fu H, Rao M, Sussel L, Rubenstein J, Qiu M (2001) Control of oligodendrocyte differentiation by the Nkx2.2 homeodomain transcription factor. Development 128:2723–2733

Ragsdale CW, Grove EA (2001) Patterning the mammalian cerebral cortex. Curr Opin Neurobiol 11:50–58

Reddy PH, Williams M, Tagle DA (1999) Recent advances in understanding the pathogenesis of Huntington's disease. Trends Neurosci 22:248–255

Redwine JM, Armstrong RC (1998) In vivo proliferation of oligodendrocyte progenitors expressing PDGFalphaR during early remyelination. J Neurobiol 37:413–428

Reynolds R, Hardy R (1997) Oligodendroglial progenitors labeled with the O4 antibody persist in the adult rat cerebral cortex in vivo. J Neurosci Res 47:455–470

Richardson WD, Pringle NP, Yu WP, Hall AC (1997) Origins of spinal cord oligodendrocytes: possible developmental and evolutionary relationships with motor neurons. Dev Neurosci 19:58–68

Ross B, Bluml S (2001) Magnetic resonance spectroscopy of the human brain. New Anat 265:54–84

Rowitch DH, S-Jacques B, Lee SMK, Flax JD, Snyder EY, McMahon AP (1999) Sonic hedgehog regulates proliferation and inhibits differentiation of CNS precursor cells. J Neurosci 19: 8954–8965

Rozental R, Srinivas M, Gökhan S, Urban M, Dermietzel R, Kessler JA, Spray DC, Mehler MF (2000) Temporal expression of neuronal connexins during hippocampal ontogeny. Brain Res Rev 32: 57–71

Roztocil T, Matter-Sadzinski L, Alliod C, Ballivet M, Matter JM (1997) NeuroM, a neural helix-loop-helix transcription factor, defines a new transition stage in neurogenesis. Development 124:3263–3272

Sakurada K, Ohshima-Sakurada M, Palmer TD, Gage FH (1999) Nurr1, an orphan nuclear receptor, is a transcriptional activator of endogenous tyrosine hydroxylase in neural progenitor cells derived from the adult brain. Development 126:4017–4026

Satow T, Bae SK, Inoue T, Inoue C, Miyoshi G, Tomita K, Bessho Y, Hashimoto N, Kageyama R (2001) The basic helix-loop-helix gene hesr2 promotes gliogenesis in mouse retina. J Neurosci 21:1265–1273

Sieradzan KA, Mann DMA (2001) The selective vulnerability of nerve cells in Huntington's disease. J Neuropathol Appl Neurobiol 27:1–21

Sun Y, Nadal-Vicens M, Misono S, Lin MZ, Zubiaga A, Hua X, Fan G, Greenberg ME (2001) Neurogenin promotes neurogenesis and inhibits glial differentiation by independent mechanisms. Cell 104:365–376

Sussel L, Marin O, Kimura S, Rubenstein JLR (1999) Loss of Nkx2.1 homeobox gene function results in a ventral to dorsal molecular respecification within the basal telencephalon: evidence for a transformation of the pallidum into the striatum. Development 126:3359–3370

Takebayashi H, Yoshida S, Sugimori M, Kosako H, Kominami R, Nakafuku M, Nabeshima Y (2000) Dynamic expression of basic helix-loop-helix Olig family members: implication of Olig2 in neuron and oligodendrocyte differentiation and identification of a new member, Olig3. Mech Dev 99:143–148

Tamamaki N, Fujimori E, Takauji R (1997) Origin and route of tangentially migrating neurons in the developing neocortical intermediate zone. J Neurosci 17:8313–8323

Tamamaki N, Sugimoto Y, Tanaka K, Takauji R (1999) Cell migration from the ganglionic eminence to the neocortex investigated by labeling nuclei with UV irradiation via a fiber-optic cable. Neurosci Res 35:241–251

Tanigaki, K, Nogaki F, Takahashi J, Tashiro K, Kurooka H, Honjo T (2001) Notch1 and Notch3 instructively restrict bFGF-responsive multipotent neural progenitor cells to an astroglial fate. Neuron 29:45–55

Tekki-Kessaris N, Woodruff R, Hall AC, Gaffield W, Kimura S, Stiles CD, Rowitch DH, Richardson WD (2001) Hedgehog-dependent oligodendrocyte lineage specification in the telencephalon. Development 128:2545–2554

Thomas JL, Spassky N, Perez Villegas EM, Olivier C, Cobos I, Goujet-Zalc C, Martinez S, Zalc B (2000) Spatiotemporal development of oligodendrocytes in the embryonic brain. J Neurosci Res 59:471–476

Tomita K, Moriyoshi K, Nakanishi S, Guillemot F, Kageyama R (2000) Mammalian achaete-scute and atonal homologs regulate neuronal versus glial fate determination in the central nervous system. EMBO J 19:5460–5472

Vonsattel J-P, Myers RH, Stevens TJ, Ferrante MS, Bird ED, Richardson EP (1985) Neuropathological classification of Huntington's disease. J Neuropathol Exp Neurol 44:559–577

Wang S, Barres BA (2000) Up a notch: instructing gliogenesis. Neuron 27:197–200

Weiss S, Dunne C, Hewson J, Wohl C, Wheatley M, Peterson AC, Reynolds BA (1996a) Multipotent CNS stem cells are present in the adult mammalian spinal cord and ventricular neuroaxis. J Neurosci 16:7599–7609

Weiss S, Reynolds BA, Vescovi AL, Morshead C, Craig CG, van der Kooy D (1996b) Is there a neural stem cell in the mammalian forebrain. Trends Neurosci 9:387–393

Wichterle H, Garcia-Verdugo JM, Herrera DG, Alvarez-Buylla A (1999) Young neurons from medial ganglion eminence disperse in adult and embryonic brain. Nat Neurosci 2:461–466

Wilson SW, Rubenstein JLR (2000) Induction and dorsoventral patterning of the telencephalon. Neuron 28:641–651

Woodruff RH, Tekki-Kessaris N, Stiles CD, Rowitch DH, Richardson WD (2001) Oligodendrocyte development in the spinal cord and telencephalon: common themes and new perspectives. Int J Dev Neurosci 19:379–385

Xu X, Cai J, Fu H, Wu R, Qi Y, Modderman G, Liu R, Qiu M (2000) Selective expression of Nkx-2.2 transcription factor in chicken oligodendrocyte progenitors and implications for the embryonic origin of oligodendrocytes. Mol Cell Biol 16:740–753

Zhou Q, Wang S, Anderson DJ (2000) Identification of a novel family of oligodendrocyte lineage-specific basic helix-loop-helix transcription factors. Neuron 25:331–343

Subject Index

Printing: Mercedes-Druck, Berlin
Binding: Stein+Lehmann, Berlin